Julian Peter Lemburg

Pricing Bermudan Options by Forward Improvement Iteration

Julian Peter Lemburg

Pricing Bermudan Options by Forward Improvement Iteration

Südwestdeutscher Verlag für Hochschulschriften

Impressum/Imprint (nur für Deutschland/only for Germany)
Bibliografische Information der Deutschen Nationalbibliothek: Die Deutsche Nationalbibliothek verzeichnet diese Publikation in der Deutschen Nationalbibliografie; detaillierte bibliografische Daten sind im Internet über http://dnb.d-nb.de abrufbar.

Alle in diesem Buch genannten Marken und Produktnamen unterliegen warenzeichen-, marken- oder patentrechtlichem Schutz bzw. sind Warenzeichen oder eingetragene Warenzeichen der jeweiligen Inhaber. Die Wiedergabe von Marken, Produktnamen, Gebrauchsnamen, Handelsnamen, Warenbezeichnungen u.s.w. in diesem Werk berechtigt auch ohne besondere Kennzeichnung nicht zu der Annahme, dass solche Namen im Sinne der Warenzeichen- und Markenschutzgesetzgebung als frei zu betrachten wären und daher von jedermann benutzt werden dürften.

Coverbild: www.ingimage.com

Verlag: Südwestdeutscher Verlag für Hochschulschriften GmbH & Co. KG
Dudweiler Landstr. 99, 66123 Saarbrücken, Deutschland
Telefon +49 681 37 20 271-1, Telefax +49 681 37 20 271-0
Email: info@svh-verlag.de

Zugl.: Kiel, Christian-Albrechts-Universität, Dissertation, 2011

Herstellung in Deutschland:
Schaltungsdienst Lange o.H.G., Berlin
Books on Demand GmbH, Norderstedt
Reha GmbH, Saarbrücken
Amazon Distribution GmbH, Leipzig
ISBN: 978-3-8381-2630-2

Imprint (only for USA, GB)
Bibliographic information published by the Deutsche Nationalbibliothek: The Deutsche Nationalbibliothek lists this publication in the Deutsche Nationalbibliografie; detailed bibliographic data are available in the Internet at http://dnb.d-nb.de.

Any brand names and product names mentioned in this book are subject to trademark, brand or patent protection and are trademarks or registered trademarks of their respective holders. The use of brand names, product names, common names, trade names, product descriptions etc. even without a particular marking in this works is in no way to be construed to mean that such names may be regarded as unrestricted in respect of trademark and brand protection legislation and could thus be used by anyone.

Cover image: www.ingimage.com

Publisher: Südwestdeutscher Verlag für Hochschulschriften GmbH & Co. KG
Dudweiler Landstr. 99, 66123 Saarbrücken, Germany
Phone +49 681 37 20 271-1, Fax +49 681 37 20 271-0
Email: info@svh-verlag.de

Printed in the U.S.A.
Printed in the U.K. by (see last page)
ISBN: 978-3-8381-2630-2

Copyright © 2011 by the author and Südwestdeutscher Verlag für Hochschulschriften GmbH & Co. KG and licensors
All rights reserved. Saarbrücken 2011

Contents

Outline   5

Preface   5

1 Setting and Definitions   9
- 1.1 Conventions 9
- 1.2 Setting 9
- 1.3 Definition and Remark: Stopping Rule and Snell Envelope 9
- 1.4 Lemma: Integrability 10
- 1.5 Theorem: Snell Envelope 10
- 1.6 Theorem: Convergence of the Snell Envelope 11
- 1.7 Theorem: Backward Dynamic Programming of the Snell Envelope 11
- 1.8 Convention 11
- 1.9 Convention: Fixed N 11
- 1.10 Definition: (Consistent) N-Suitable Family of Stopping Rules 12
- 1.11 Definition: Optimal 12
- 1.12 Lemma: Consistent, Optimal 13
- 1.13 Definition and Theorem:
First and Last Optimal N-Suitable Family of Stopping Rules 13

2 Forward Improvement Iteration Algorithm   15
- 2.1 Definitions 15
- 2.2 Definition:
N-suitable Adapted Random Set and Generated Family 16
- 2.3 Definition and Theorem: Corresponding 16
- 2.4 Remark: Corresponding 18
- 2.5 Definition: Corresponding To 18
- 2.6 Definition and Remark: Essential 18
- 2.7 Lemma: Elements of $\mathcal{T}$ compared 19
- 2.8 The Window Parameter 20
- 2.9 Input for FII Algorithm 20
- 2.10 Remark: Input for FII Algorithm 20
- 2.11 Output of FII Algorithm 21
- 2.12 Remarks: Output of FII Algorithm 21
- 2.13 Performing FII Algorithm 22
- 2.14 Remarks: Performing FII Algorithm 23
- 2.15 Comparison with the Literature 23
- 2.16 Lemma: Elementary Equations and Inequalities 24
- 2.17 Lemma: Existence of Improvers 25
- 2.18 Definition: Improver 25
- 2.19 Setting of [Irl80] included 26

2.20	Remark: Infinite Case Includes Finite Case	26

3 Finite Time Horizon · 27

3.1	Theorem: Equations and Inequalities	27
3.2	Lemma	30
3.3	Theorem	30
3.4	Conclusion	31
3.5	Example	32
3.6	Lemma	34
3.7	Theorem	35
3.8	Theorem	35
3.9	Remark	37
3.10	Theorem	37
3.11	Lemma	38
3.12	Lemma	38
3.13	Conclusion	39
3.14	Theorem: Optimal Algorithm Termination for $C^{(0)} \equiv \Omega$	39
3.15	Definition and Remark	41
3.16	Example	41
3.17	Generalized Example	42
3.18	Theorem: Algorithm Termination	42
3.19	Example	43
3.20	Lemma	44
3.21	Theorem: Influence of κ	44

4 Different Finite Time Horizons Compared · 47

4.1	Idea	47
4.2	Example	47
4.3	Lemma	48
4.4	Lemma	49
4.5	Theorem	50

5 General Case · 51

5.1	Theorem from [Irl80] for $\kappa \equiv 1$	51
5.2	General Assumption	52
5.3	Main Theorem	52
5.4	Delimitating Example	52
5.5	Remark: Limites	53
5.6	Remarks	53
5.7	Lemma	53
5.8	Remarks	54
5.9	Definitions	54
5.10	Lemma: Optimality	55
5.11	Lemma: One Step Improvement	55
5.12	Theorem	61
5.13	Conclusion	63
5.14	Conclusion	64
5.15	Remark	64
5.16	Outlook	64

6 Markovian Case   65
- 6.1 Setting 65
- 6.2 Time-Independent Pay-Off 65
  - 6.2.1 Connection to the General Case 66
- 6.3 Time-Dependent Pay-Off 67
  - 6.3.1 Finite Horizon 67
  - 6.3.2 Linear Costs 67
  - 6.3.3 Constant Discounting 68
- 6.4 Path-Dependent Costs 68
- 6.5 Performing the Algorithmic Step 69
- 6.6 Outlook 70

7 Markovian Case with Random Discounting   71
- 7.1 Motivation 71
- 7.2 Setting 72
- 7.3 Remarks 73
- 7.4 Lemma: One-Step Improvement 73
- 7.5 Example: Insufficient Improvement 74
- 7.6 Lemma: Partial Improving I 75
- 7.7 Lemma: Partial Improving II 76
- 7.8 Example: Necessities for Partial Improving II 76
- 7.9 Proof of Lemma 7.4 (page 73) 78
- 7.10 Remark 80
- 7.11 Theorem 81

8 Markovian Case with Random Discounting and Finite State Space   85
- 8.1 Setting 85
- 8.2 Theorem 85
- 8.3 Corollar 86
- 8.4 FII by Solving Linear Equations 87

9 Numerical Examples   89
- 9.1 Setting 89
- 9.2 Results for $\kappa \equiv \{1\}$ 90
- 9.3 Influence of κ 92

10 Monotone Case   95
- 10.1 Definition 95
- 10.2 Remark: Appearance in the above Model 96
- 10.3 Definition and Remark:
  m-monotone and m-stages look-ahead rules 96
- 10.4 Lemma: m-monotone 97
- 10.5 Conclusion 97
- 10.6 Counter-Example: m-monotone 98
- 10.7 Lemma 99
- 10.8 Theorem 99
- 10.9 Lemma 100
- 10.10 Markov Case 100

11 No-Information Version of the Best Choice Problem with Random Population Size 101

- 11.1 General Setting 101
- 11.2 Special Setting 102
- 11.3 Definitions and Remarks 102
- 11.4 Lemma 104
- 11.5 Lemma 104
- 11.6 Lemma 105
- 11.7 Lemma 105
- 11.8 Lemma 105
- 11.9 Lemma 105
- 11.10 Theorem 106
- 11.11 Theorem 107
- 11.12 Setting Continued 109
- 11.13 Theorem 109
- 11.14 Lemma 110
- 11.15 Lemma: Optimality 110
- 11.16 Lemma: Problem-Independent Positive Values of a 111
- 11.17 Conclusion about a and a_κ 112
- 11.18 Remark 112
- 11.19 Lemma 112
- 11.20 Conclusion: Reasonable Minimal Value for d 112
- 11.21 Lemma: Bounded Values 112
- 11.22 Conclusion: Bounded Values 113
- 11.23 Lemma: Limit Value Zero 113
- 11.24 Theorem 113
- 11.25 Theorem 114
- 11.26 Lemma 114
- 11.27 Theorem: Deterministic Population Size 115
- 11.28 Conclusion 117
- 11.29 Remark 118
- 11.30 Conclusion 118

Bibliography   **119**

Outline

After a detailed introduction we will show necessary basics in Chapter 1 (page 9). The algorithm with all its extensions will be presented in Chapter 2 (page 15). In Chapter 3 (page 27) and Chapter 4 (page 47) its application given finite time horizon is done and in Chapter 5 (page 51) its application given arbitrary time horizon. The algorithm is transferred to the Markovian case in Chapter 6 (page 65), Chapter 7 (page 71) und Chapter 8 (page 85). Numerical examples follow in Chapter 9 (page 89). The application in the "monotone case" is shown in Chapter 10 (page 95) and some advantages of the algorithm are shown for a best choice problem in Chapter 11 (page 101).

Preface

At a stock exchange, not only stocks change hands, but also bets on stocks. These are called financial products or derivatives. Quite common is the following bet: You pay a certain price and receive the right to say stop at one of some given time points and then receive a payoff directly depending on the current price of one or more other financial products, often plain stocks. This bet is called a Bermudan option.

For example the payment is obtained as the amount, by which the actual price of a particular share exceeds a predetermined value (called strike price). This is often called Bermudan call option. Ideally it is to stop on the date on which the payment has its highest value. – But at each time point it is unclear whether tomorrow would be an even better time point. It is therefore necessary to find a rule, when you should stop. The highest expected payoff that a rule can bring is accordingly the fair price of the option, but how is it determined or approximated?

Pricing
Pricing of Bermudan style derivatives on a high dimensional system of underlyings is considered an enduring problem for the last years. Prices for such high dimensional options are difficult, if not impossible, to compute by standard partial differential equation (PDE) methods. For high dimensional European options a good alternative to PDEs is Monte Carlo simulation.[1] Nevertheless, for American options, Monte Carlo simulation is more complicated since the (optimal) exercise boundary is usually unknown.

Various Monte Carlo algorithms for pricing American and Bermudan options have been developed and described in the literature. For a detailed and general overview we refer to [Gla03] and [Put94] and the references therein. Many of these algorithms are related to backward

---
[1] The name goes back to John von Neuman (1903-1957), who used "Monte Carlo" in 1946 as a code name for a secret project, wherein he used these techniques. The name refers to the Monte Carlo casino in Monaco.

dynamic programming which comes down to a recursive representation of the Snell envelope. They require the evaluation of high order nestings of conditional expectations.

Therefore Monte Carlo estimators for regression functions, which do not run into explosive cost when nested several times, have been proposed by several authors.

As an alternative to backward dynamic programming, one may search for a suitable parametric family of exercise boundaries and then maximize the solutions of the corresponding family of boundary value problems over the parameters.

Policy Iteration – Approximations Already After the First Iteration

Another alternative to solving the backward dynamic program recursively are policy iterations. Their main advantage is that they mend one of the main drawbacks of the backward scheme: Assume exercise can take place at one out of k time instances. To obtain the value of the optimal stopping problem via backward dynamic programming, we have to calculate nested conditional expectations of the order k. The kth nested conditional expectation has to be evaluated before any approximation of the time 0 value is possible. This prevents the use of plain Monte-Carlo simulations for approximating the conditional expectations and requires more complicated approximation procedures for these quantities. In contrast, policy iteration algorithms yield already after each iteration step an improved approximation of the Snell envelope which ranges over all exercise dates. This allows to calculate approximations of the Snell envelope of a Markovian process via a plain Monte-Carlo simulation.

Forward Improvement Iteration

A classical method of policy iteration, which is well-known in dynamic programming, was developed by Howard in [How60].[2] Adapting this method to problems of optimal stopping yields a method first published by Irle in [Irl80] and later called "forward improvement iteration" (FII) (see [Irl06] and [Irl09]). The mathematical model of optimal stopping problems is considerably less involved than the general model of dynamic programming. Hence Howard's method has a simple form and the optimal stopping problem can be solved by the resulting algorithm without any restrictions on the underlying state space and distribution of the process. We will discuss this algorithm further in this book.

In the method of backward induction we have to compare the actual pay-off with the pay-off due to *optimal* continuation first at the last stage, then at the second-last stage and so forth backwards in time. Instead of that the forward improvement algorithm starts with a sequence of attainable sets and compares the pay-off at each stage with the pay-off due to stopping at the next attainable set. Repeating this, one arrives at a sequence of sets which characterizes an optimal stopping rule. By using stopping rules, Bender, Kolodko and Shoenmarkers have developed a similar approach in [KS06] and [BS06].[3] Some random sets appear in [KS06, part 6] as additional parameters. We will show below that these papers describe in fact the same algorithm and the same idea as already published in [Irl80], just seen from another viewpoint. *We will show the existence of a one to one correspondence between the above mentioned attainable sets, the mentioned sequences of stopping rules and the mentioned random sets.*

The central result in the above mentioned papers ([Irl80], [Irl06], [Irl09], [KS06] and [BS06]) is an

---

[2]See the explanations in [BKS08, Introduction], [KS06, Part 1] and [Irl80, page 179].
[3]See the abstracts in [BS06] and [BKS08].

iterative construction of the Snell envelope via a sequence of attainable sets and a corresponding sequence of stopping rules which increases to the (first) optimal stopping rule (and equivalently to a sequence of sets and a sequence of random sets), corresponding to the optimal stopping rule. In each iteration step we improve a whole family of stopping rules, attainable sets and random sets. This family is denoted with (τ_i, C_i, Θ_i) where i runs through the set of exercise dates. τ_i is the stopping rule for the Bermudan which is not exercised before date i. The thus obtained sequence of stopping families naturally induces a non-decreasing sequence of lower approximations of the Snell envelope. When looking at a model with finitely many exercise dates, the sequence even coincides with the Snell envelope after finitely many steps: The number of necessary steps is at most the number of exercise dates.

A canonical suboptimal exercise policy for Bermudan swaptions is: exercise as soon as the cash-flow dominates all the Europeans ahead. This is the first iteration step of the FII algorithm (when starting with the largest possible attainable set). The algorithm is just iterating this policy. Schoenmakers states that even the first iteration is usually "already not far from optimal". Due to its special nature the FII algorithm gives usually good results with only very few iterations.[4]

Continuous Time
The whole analysis is based on a discrete set of exercise dates, but since a continuous time American option may be approximated by a Bermudan option with a fine grid of exercise dates, one may in principle apply the method as well.

Infinite Time Horizon
In [KS06] it was depicted that the therein shown algorithm for a finite set of exercise dates can be used for an perpetual discrete optimal stopping problem by approximation through instances with a finite set of exercise dates. But this is not necessary! Irle highlighted already in [Irl80] that the procedure has the advantage that problems with infinite time horizon can be treated directly and easily. Clearly, most of the proofs in [KS06] and [BS06] are based on backward induction which cannot be transferred to problems of infinite time-horizon. Other techniques are necessary, the framework for these was developed in [Irl80] and further developed and explained in [Irl06] and [Irl09].

Window Parameter
Irle used a window parameter implicitly set to one for the algorithm when describing it in [Irl80]. Bender and Schoenmakers used a similar parameter, which in their work [BS06] is set to the number of time-steps. In [KS06] Kolodko and Schoenmakers explicitly mentioned this window parameter, called it κ and examined its influence for the finite case on the *first* iteration step. *We will show for all iteration steps that in the finite case a larger κ is at least as good as using a smaller one. We will show empirically that a small κ around 4 increases the speed of the algorithm (when using linear equations). We will examine the influence of κ for the infinite case and show that the choice does not change the convergence results formerly shown by Irle.* The window parameter is roughly spoken the number of time points we look ahead in each iteration step to improve our results:

For each iteration of the algorithm step we fix some window parameter k. Let l be the minimum of k and the number of remaining time points ahead. We look l time points ahead in this

---
[4]See [KS06, Introduction] and [Irl09, Introduction].

manner: We compare the current pay-off with the expected pay-off when we do not use the next 1, 2, 3, ... l exercise possibilities and stop then according to the best stopping rule we have calculated until now. The window parameter k is the maximal number of time points we look ahead. It may be chosen differently in each step and may even depend on the foregoing iteration step.

Scenario Selection Method
In principle, the forward improvement iteration algorithm can be started with a very simple input policy like "exercise immediately". Given today's computer power however, more than one degree of nesting conditional expectations is virtually impossible (when using Monte Carlo). This means that in practice only one step of the algorithm can be carried out.[5] Therefore, the choice of the input stopping family is important.

The basic idea is as follows: Suppose the holder of the option has some pre-information, for example he knows good closed form approximations of the price for the corresponding European options due to their availability in the literature for practically all relevant options. So the investor may rule out some scenarios, at which an optimal strategy cannot exercise, by the pre-information. Then the policy improvement is run only at the remaining time points. The set of remaining time points depends on the state of the underlying system, thus we do not simply reduce to an other option with less exercise dates.

This scenario selection method has already been described in [Irl80] and the subsequent papers of Irle and was depicted in [BKS08, part 3] in a slightly different formulation. *We will show their equivalence.*

---

[5]See [BKS08, Introduction].

1 Setting and Definitions

1.1 Conventions

- If we introduce a new term, we write it in bold letters.
- If we introduce a new variable for permanent use in this book, we write it in bold letters.
- We write just a for families $(a_i)_{i \in I}$ with index set I, wherever I is obvious by the context.

1.2 Setting

Let us consider the following general situation for a problem of optimal stopping. We start with some probability space $(\mathbf{\Omega}, \mathbf{\mathcal{A}}, \mathbf{P})$ and a filtration $(\mathbf{\mathcal{A}_n})_{0 \leqslant n < \infty}$ with $\mathcal{A}_0$ being trivial and an adapted real-valued stochastic process $(\mathbf{X_n})_{0 \leqslant n < \infty}$, such that X_n is integrable for all $n \in \mathbb{N}_0$ and $\mathbf{X_\infty} = \limsup_{n \to \infty, k \geqslant n} X_k$. Define $\mathbf{X^N} := (X_n)_{0 \leqslant n \leqslant N}$ for all $N \in \mathbb{N}_0 \cup \{\infty\}$.

1.3 Definition and Remark: Stopping Rule and Snell Envelope

A **stopping rule** is a mapping $\tau : \Omega \longrightarrow \mathbb{N}_0 \cup \{\infty\}$ satisfying $\{\tau = n\} \in \mathcal{A}_n$ for all $n \in \mathbb{N}_0$. Let $\mathcal{S}$ denote the set of all stopping rules. Some subsets of the set of all stopping rules will be of special interest. So define for all $n \in \mathbb{N}_0$

$$\mathcal{S}_n^\infty := \{\tau \in \mathcal{S} \,;\, n \leqslant \tau \leqslant \infty, E\left(X_\tau^-\right) < \infty\},$$
$$\mathcal{S}_n^b := \{\tau \in \mathcal{S} \,;\, \exists N \in \mathbb{N}_0 \ n \leqslant \tau \leqslant N\},$$
$$\mathcal{S}_n^* := \{\tau \in \mathcal{S} \,;\, n \leqslant \tau < \infty, E\left(X_\tau^+\right) < \infty \text{ or } E\left(X_\tau^-\right) < \infty\},$$
$$\overline{\mathcal{S}}_n := \{\tau \in \mathcal{S} \,;\, n \leqslant \tau < \infty, E\left(X_\tau^-\right) < \infty\},$$

and for all $n, N \in \mathbb{N}_0$ with $n \leqslant N$ define

$$\mathcal{S}_n^N := \{\tau \in \mathcal{S} \,;\, n \leqslant \tau \leqslant N\}.$$

The definition of $\mathcal{S}_n^\infty$ is already done above and slightly different.

We have

(1.3.1)
$$\mathcal{S}_n^b = \bigcup_{n \in \mathbb{N}_0} \mathcal{S}_n^N.$$

Distinguishing different sets of stopping rules in the finite case (meaning N is finite) is not necessary due to the fact that for all $N \in \mathbb{N}_0$, $n \in \mathbb{N}_0$ with $n \leqslant N$ and for all $\tau \in \mathcal{S}_n^N$ the random variable X_τ is integrable, as we will show in Lemma 1.4 (page 10) below.

Furthermore define for all $N \in \mathbb{N}_0 \cup \{\infty\}$ the process $Y_N^* = (Y_{N,n}^*)_{0 \leq n \leq N}$ by

$$Y_{N,n}^* := \operatorname{esssup} \left\{ \mathrm{E}\left(X_\tau | \mathcal{A}_n\right) \ ; \ \tau \in \mathcal{S}_n^N \right\} \text{ for all } n \in \mathbb{N}_0 \text{ with } n \leq N,$$

often called **Snell envelope of X^N** named after James Laurie Snell (born January 15, 1915), a student of Joseph Leo Doob (1910-2004), due to the article [Sne52] he published in 1952.

1.4 Lemma: Integrability

(1.4.1) X_τ is integrable for all $\tau \in \mathcal{S}_0^b$,

(1.4.2) $\sup_{m \in \mathbb{N}_0} |X_{\sigma_m}|$ is integrable for all $\sigma \in \left(\mathcal{S}_n^N\right)^{\mathbb{N}}$ for all $N, n \in \mathbb{N}_0$ with $n \leq N < \infty$

Proof: Let $n \in \mathbb{N}_0$. Due to Formula (1.3.1) (page 9) it suffices to concentrate on $\mathcal{S}_n^N$ for some $N \in \mathbb{N}_0$ instead of examining $\mathcal{S}_0^b$ to prove (1.4.1). Hence consider $N \in \mathbb{N}_0$ with $n \leq N < \infty$. For all $l \in \mathbb{N}_0$ with $l \leq N$ we have $|X_l|$ integrable by Setting 1.2 (page 9). Define $W := \sum_{l=0}^{N} |X_l|$. As a sum of integrable random variables W is integrable. Since

(1.4.3) $$|X_\tau| \leq \max\{|X_l| \ ; \ 0 \leq l \leq N\} \leq \sum_{l=0}^{N} |X_l| = W \text{ for all } \tau \in \mathcal{S}_n^N,$$

we have the integrability in (1.4.1).

For all $m \in \mathbb{N}_0$ we have $|X_{\sigma_m}| \leq W$ by (1.4.3) and thus

$$0 \leq \sup\{|X_{\sigma_m}| \ ; \ m \in \mathbb{N}_0\} \leq W,$$

hence we have the integrability in (1.4.2). □

1.5 Theorem: Snell Envelope

We have for all $n \in \mathbb{N}_0$

$$\begin{aligned} Y_{\infty,n}^* &= \operatorname{esssup}\{\mathrm{E}\left(X_\tau | \mathcal{A}_n\right) \ ; \ \tau \in \mathcal{S}_n^\infty\} \\ &= \operatorname{esssup}\{\mathrm{E}\left(X_\tau | \mathcal{A}_n\right) \ ; \ \tau \in \mathcal{S}_n^*\} \\ &= \operatorname{esssup}\{\mathrm{E}\left(X_\tau | \mathcal{A}_n\right) \ ; \ \tau \in \overline{\mathcal{S}}_n\} \end{aligned}$$

and these essential suprema are not enlarged by allowing "randomized" stopping rules. For all $N \in \mathbb{N}_0$ and under certain assumptions also for $N = \infty$ the process Y_N^* is the smallest X^N-dominating supermartingale.

Proof: See for example [CRS71, pp. 42, 63, 78 for definitions and pp. 63, 80, 81 for results] and [CRS71, p.111,th.5.3]. Beware the different naming of the variables: $\overline{\mathcal{S}}_n \triangleq C_n^\infty$, $\mathcal{S}_n^\infty \triangleq \overline{C}_n$, $\mathcal{S}_n^* \triangleq C_n^*$. □

1.6 Theorem: Convergence of the Snell Envelope

We have convergence of the Snell Envelope in the sense of

$$Y^*_{\infty,n} = \lim_{N \to \infty} Y^*_{N,n} = \operatorname{esssup}\{E(X_\tau|\mathcal{A}_n) \; ; \; \tau \in \mathcal{S}^b_n\} \text{ for all } n \in \mathbb{N}_0,$$

if one of the following conditions is true:

(a) $X_n \geq 0$ for all $n \in \mathbb{N}_0 \cup \{\infty\}$,
(b) $(X_n)^- \leq W$ for all $n \in \mathbb{N}_0 \cup \{\infty\}$ for some integrable random variable $W \geq 0$,
(c) $((X_n)^-)_{n \in \mathbb{N}_0}$ is uniformly integrable
(d) $\lim_{n \to \infty} \int_{\{\tau > n\}} (X_n)^- dP = 0$ for all $\tau \in \overline{\mathcal{S}}_0$,
(e) $\lim_{n \to \infty} \inf_{k \geq n} \int_{\{\tau > k\}} (\lim_{N \to \infty} Y^*_{N,k})^- dP = 0$ for all $\tau \in \overline{\mathcal{S}}_0$,

Remark: We have (a) $\Longrightarrow$ (b), (b) $\Longrightarrow$ (c), (c) $\Longrightarrow$ (d), (d) $\Longrightarrow$ (e).

Proof: A proof can be found in [CRS71, p. 68-70]. □

1.7 Theorem: Backward Dynamic Programming of the Snell Envelope

For finite N the Snell envelope Y^*_N can be constructed by backward dynamic programming as follows: At the last exercise date N define

$$Y^*_{N,N} := X_N,$$

and for all $n \in \mathbb{N}_0$ with $n < N$ and $Y^*_{N,n+1}$ already defined set

$$Y^*_{N,n} := \max\{X_n, E(Y^*_{N,n+1}|\mathcal{A}_n)\}.$$

Proof: A proof can be found in [Irl02, pp. 92-95]. □

1.8 Convention

For being able to write about the finite and infinite case simultaneously we often use the expression

$$"n < N + 1",$$

by assuming $\infty + 1 := \infty$ it is equivalent to

$$"n \leq N \text{ for finite } N \text{ and } n < \infty \text{ for } N = \infty".$$

1.9 Convention: Fixed N

Fix some $\boldsymbol{N} \in \mathbb{N}_0 \cup \{\infty\}$. Nearly all of the following variables depend on N. For brevity we will often abstain from showing this dependency by writing N as an index. So we write for example Y^*_n for $Y^*_{N,n}$.

1.10 Definition: (Consistent) N-Suitable Family of Stopping Rules

An **N-suitable family of stopping rules** is a family

$$\tau = (\tau_n)_{0 \leq n < N+1} \text{ with } \tau_n \in \mathcal{S}_n^N \text{ for all } n \in \mathbb{N}_0 \text{ with } n < N+1.$$

Using a definition introduced in [BKS08, def. 3.1]
it is called **consistent** iff

$$\{\tau_n > n\} \subseteq \{\tau_n = \tau_{n+1}\} \text{ for all } n \in \mathbb{N}_0 \text{ with } n < N,$$

hence (since $\tau_{n+1} \geq n+1$) iff

$$\{\tau_n > n\} = \{\tau_n = \tau_{n+1}\} \text{ for all } n \in \mathbb{N}_0 \text{ with } n < N.$$

Define the set of all consistent N-suitable families of stopping rules by

$$\mathcal{T} := \left\{(\tau_n)_{0 \leq n < N+1} \ ; \ \forall \, 0 \leq n < N+1 \ \tau_n \in \mathcal{S}_n^N, \forall \, 0 \leq n < N \ \{\tau_n > n\} \subseteq \{\tau_n = \tau_{n+1}\}\right\}.$$

1.11 Definition: Optimal

Consider $\hat{\mathcal{S}} \subseteq \mathcal{S}$ and $n \in \mathbb{N}_0$ with $n < N+1$.

Any $\tau \in \mathcal{S}$ is called **optimal in $\hat{\mathcal{S}} \cap \mathcal{S}_n^N$** iff

$$\tau \in \hat{\mathcal{S}} \cap \mathcal{S}_n^N \text{ and } \text{E}(X_\tau | \mathcal{A}_n) = \operatorname*{esssup}\left\{\text{E}(X_\sigma | \mathcal{A}_n) \ ; \ \sigma \in \hat{\mathcal{S}} \cap \mathcal{S}_n^N\right\}.$$

Hence a $\tau \in \mathcal{S}$ is optimal in $\mathcal{S}_n^N$ iff

$$\tau \in \mathcal{S}_n^N \text{ and } \text{E}(X_\tau | \mathcal{A}_n) = Y_n^*.$$

Any $\tau \in \mathcal{T}$ is called **optimal in $\hat{\mathcal{S}}$** iff

$$\tau_n \text{ is optimal in } \hat{\mathcal{S}} \cap \mathcal{S}_n^N \text{ for all } n \in \mathbb{N}_0 \text{ with } n < N+1,$$

and called **optimal** iff

$$\tau_n \text{ is optimal in } \mathcal{S}_n^N \text{ for all } n \in \mathbb{N}_0 \text{ with } n < N+1.$$

Hence not mentioning $\hat{\mathcal{S}}$ is a shortcut if $\hat{\mathcal{S}} = \mathcal{S}$.

1.12 Lemma: Consistent, Optimal

- Each optimal N-suitable family of stopping rules is consistent.
- For all $\tau \in \mathcal{T}$ we have
$$\tau_n \leq \tau_{n+1}$$
for all $n \in \mathbb{N}_0$ with $n < N$.
- For all $\tau \in \mathcal{T}$ and $\omega \in \Omega$ the sequence $(\tau_n(\omega))_{n \in \mathbb{N}_0}$ is non-decreasing and piecewise constant in the following sense. For all $n \in \mathbb{N}_0$ with $n < N$ we have:
If $\tau_n(\omega) < \infty$ then there is a minimal $m = m(\omega) \in \mathbb{N}_0$ such that we have
$$\tau_n(\omega) = \tau_{n+k}(\omega) = \tau_{n+m}(\omega) = n + m \text{ for all } k \in \mathbb{N}_0 \text{ with } k < m.$$
If $\tau_n(\omega) = \infty$ then $\tau_{n+k}(\omega) = \infty$ for all $k \in \mathbb{N}_0$.

Proof: This follows easily by induction from the definition of $\mathcal{T}$ in Definition 1.11 (page 12). □

1.13 Definition and Theorem: First and Last Optimal N-Suitable Family of Stopping Rules

Define
$$\check{\tau}^* := (\check{\tau}_n^*)_{0 \leq n < N+1} := \left(\inf \left\{ n \; ; \; n \leq i \leq N, Y_i^* \leq X_i \right\} \right)_{0 \leq n < N+1}$$
$$= \left(\inf \left\{ n \; ; \; n \leq i \leq N, Y_i^* = X_i \right\} \right)_{0 \leq n < N+1},$$
$$\hat{\tau}^* := (\hat{\tau}_n^*)_{0 \leq n < N+1} := \left(N \wedge \inf \left\{ n \; ; \; n \leq i \leq N, \mathrm{E}\left(Y_{i+1}^* \big| \mathcal{A}_i\right) < X_i \right\} \right)_{0 \leq n < N+1}.$$

- Consider finite N. Then $\check{\tau}^*$ is the **first** and $\hat{\tau}^*$ the **last** optimal consistent N-suitable family of stopping rules: For each optimal consistent N-suitable family of stopping rules $\tau = (\tau_n)_{0 \leq n < N+1} \in \mathcal{T}$ and for all $n \in \mathbb{N}_0$ with $n \leq N$ we have
$$\check{\tau}_n^* \leq \tau_n \leq \hat{\tau}_n^*.$$
For all $n \in \mathbb{N}_0$ with $n \leq N$ and ρ being optimal in $\mathcal{S}_n^N$ we have
$$\check{\tau}_n^* \leq \rho \leq \hat{\tau}_n^*$$

- Consider infinite N, $\check{\tau}_n^*$ a.s. finite for all $n \in \mathbb{N}_0$ and $E(\sup_{n \in \mathbb{N}_0} X_n^+) < \infty$, then $\check{\tau}^*$ is the **first** optimal consistent N-suitable family of stopping rules.

Proof: See for example [Nev75, pp. 120-129] and [BKS08, part 2]. □

2 Forward Improvement Iteration Algorithm

The following algorithm, whose conception goes back to Howard's policy improvement of dynamic programming [How60], was treated by Irle in [Irl80], [Irl06] and [Irl09] for any N and therein called *forward improvement iteration* (FII). Schoenmakers et al. treated a similar algorithm for finite N in [BS06], [KS06] and [BKS06] without naming it explicitly. They introduced a window parameter κ equal throughout all iteration steps in their papers. Irle used in his papers implicitly the window parameter one. We will assume herein a window parameter function, arbitrary chosen in each iteration step. We will call the algorithm, whenever mentioned herein, "forward improvement iteration algorithm" and it will be presented in this section.

In Definition 2.1 (page 15) and Definition 2.2 (page 16) we will define some sets herein called $\mathcal{C}$ and $\mathcal{Z}$. The first is used with the same name intensively in the papers of Irle. The latter is implicitly introduced and used in [BKS08, def. 3.1 ff.]. There is an interesting connection between these and $\mathcal{T}$, not mentioned therein, which we will call *corresponding* and examine it in Definition 2.3 (page 16) and Remark 2.4 (page 18). In Definition 2.6 (page 18) we define the term *essential* and examine it in Lemma 2.7 (page 19).

In Part 2.8 (page 20) we will introduce the above mentioned window parameter in detail. In Part 2.9 (page 20) and Remark 2.10 (page 20) we discuss the other input variables of the algorithm, the output is examined in Part 2.11 (page 21) and Remark 2.12 (page 21). In Algorithm 2.13 (page 22) we present the algorithm, and we state that in each iteration level the algorithm gives corresponding and convergent values in Remark 2.14 (page 23).

In Part 2.15 (page 23) we will compare the here stated algorithm with the similar algorithms stated before in the literature. In Lemma 2.16 (page 24) we state some elementary equations and inequalities. We define *improver*s in Definition 2.18 (page 25) and show their existence in Lemma 2.17 (page 25).

We finish the chapter with the proof in Part 2.19 (page 26) that the setting of the papers of Irle is included in the setting of this chapter and that whenever we are talking about the infinite case the finite case is included in a natural way in Remark 2.20 (page 26).

Remember that we fixed some N with $0 \leqslant N \leqslant \infty$ in Convention 1.9 (page 11).

2.1 Definitions

Define
$$\mathcal{C} := \{(C_n)_{0 \leqslant n \leqslant N} \ ; \ C_N = \Omega, \forall \, 0 \leqslant n < N \ \ C_n \in \mathcal{A}_n\},$$
and for all $C \in \mathcal{C}$
$$\mathcal{S}(C) := \{\tau \in \mathcal{S} \ ; \ \forall \, 0 \leqslant n < N \ \ \{\tau = n\} \subseteq C_n\}.$$

2.2 Definition:
N-suitable Adapted Random Set and Generated Family

A random set $A : \Omega \longrightarrow \mathcal{P}ot(\{0, ..., N\})$ such that

$N \in A$ a.s. and
for all $n \in \mathbb{N}_0$ with $n < N$ the function $1_A(n)$ is $\mathcal{A}_n$-adapted resp. $\{\omega \in \Omega \,;\, n \in A(\omega)\} \in \mathcal{A}_n$

is called an **N-suitable adapted random set**.

For every N-suitable adapted random set Θ_0 define

$$\Theta := (\Theta_n)_{0 \leqslant n < N+1} := (\Theta_0 \cap \{j \,;\, n \leqslant j \leqslant N\})_{0 \leqslant n < N+1},$$

called **the family generated by Θ_0**, it is a family of N-suitable adapted random sets.

Define $\mathcal{Z}$ as the set of the families generated by all N-suitable adapted random sets.

2.3 Definition and Theorem: Corresponding

A striking difference between the papers outlining the herein stated algorithm before ([Irl80], [Irl06], [Irl09], [BS06], [KS06] and [BKS06]) is that each of them puts the focus on only one of the sets $\mathcal{C}$, $\mathcal{Z}$ and $\mathcal{T}$ (see Definition 1.10 (page 12)), and does not explain the connection between them. Thereby it is not easy to see the similarities and differences of the papers.

We will show now that there is a certain one-to-one correspondence between these sets, which conserves all important properties. We will describe below the correspondence in detail by specifying adequate mappings.

We will say that elements of $\mathcal{C}$, $\mathcal{Z}$ and $\mathcal{T}$ are **corresponding**, if one can be constructed from the other by applying one of these mappings.

First we show that there is a mapping from $\mathcal{C}$ to $\mathcal{Z}$ and another one from $\mathcal{Z}$ to $\mathcal{C}$ so that their composition in both orders is the identity on $\mathcal{C}$ or $\mathcal{Z}$ respectively.

- Consider some $C \in \mathcal{C}$. Define

$$\Theta_0 \text{ by } \Theta_0(\omega) := \{j \,;\, 0 \leqslant j \leqslant N, \omega \in C_j\} \text{ for all } \omega \in \Omega$$

  or equivalently

$$\Theta_0 := \{j \,;\, 0 \leqslant j \leqslant N, 1_{C_j} = 1\}$$

  and further for all $n \in \mathbb{N}_0$ with $n < N + 1$

$$\Theta_n := \Theta_0 \cap \{j \,;\, n \leqslant j \leqslant N\}$$
$$= \{j \,;\, n \leqslant j \leqslant N, 1_{C_j} = 1\}.$$

  Then we have $\Theta \in \mathcal{Z}$.

- Consider some $\Theta \in \mathcal{Z}$. Define
$$C := (C_n)_{0 \leq n \leq N}$$
by
$$C_n := \{\omega \in \Omega \,;\, n \in \Theta_0(\omega)\}$$
$$= \{\omega \in \Omega \,;\, n \in \Theta_n(\omega)\} \text{ for all } n \in \mathbb{N}_0 \text{ with } n \leq N.$$
Then we have $C \in \mathcal{C}$.

- Now we show that their composition in both orders is the identity. Consider some $C \in \mathcal{C}$. Then we have for all $n \in \mathbb{N}_0$ with $n \leq N$
$$\{\omega \in \Omega \,;\, n \in \{j \,;\, 0 \leq j \leq N, \omega \in C_j\}\} = \{\omega \in \Omega \,;\, \omega \in C_n\} = C_n.$$
Conversely consider some $\Theta \in \mathcal{Z}$. Then we have for all $n \in \mathbb{N}_0$ with $n < N+1$ and $\omega \in \Omega$
$$\{j \,;\, n \leq j \leq N, \omega \in \{\omega \in \Omega \,;\, j \in \Theta_j(\omega)\}\} = \{j \,;\, n \leq j \leq N, j \in \Theta_0(\omega)\} = \Theta_n(\omega).$$

Next we show that there is a mapping from $\mathcal{C}$ to $\mathcal{T}$ and another one from $\mathcal{T}$ to $\mathcal{C}$ so that their composition in both orders is the identity on $\mathcal{C}$ or $\mathcal{T}$ respectively.

- Consider $C \in \mathcal{C}$. Define for all $n \in \mathbb{N}_0$ with $n < N+1$
$$\tau_n := \inf\{p \,;\, n \leq p \leq N \wedge \mathbf{1}_{C_p} = 1\}.$$
Then we have $\tau \in \mathcal{T}$.

- Consider $\tau \in \mathcal{T}$. Define $C_N := \Omega$ and for all $n \in \mathbb{N}_0$ with $n < N$
$$C_n := \{\tau_n = n\}.$$
Then we have $C \in \mathcal{C}$.

- Now we show that their composition in both orders is the identity. Consider some $C \in \mathcal{C}$. We have $C_N = \Omega$ and for all $n \in \mathbb{N}_0$ with $n < N$
$$\{n = \inf\{p \,;\, n \leq p \leq N \wedge \mathbf{1}_{C_p} = 1\}\}$$
$$= \{\omega \in \Omega \,;\, n = \inf\{p \,;\, n \leq p \leq N \wedge \omega \in C_p\}\}$$
$$= C_n.$$
Conversely consider $\tau \in \mathcal{T}$. Let $n \in \mathbb{N}_0$ with $n < N+1$. Let $\omega \in \Omega$. If $N = \infty$ we may have $\tau_n(\omega) = \infty$. Then it follows by definition of $\mathcal{T}$, that $\tau_p(\omega) = \infty$ for all $p \in \mathbb{N}$ with $n \leq p$, hence
$$\inf\{p \,;\, n \leq p \leq N \wedge \mathbf{1}_{\{\tau_p = p\}} = 1\}(\omega) = \infty = \tau_n(\omega).$$
Otherwise there is due to Lemma 1.12 (page 13) a minimal $m \in \mathbb{N}_0$ so that we have
$$\tau_n(\omega) = \tau_{n+k}(\omega) = \tau_{n+m}(\omega) = n + m$$
for all $k \in \mathbb{N}_0$ with $k < m$, hence
$$\tau_n(\omega) = \tau_{n+m}(\omega) = n + m = \inf\{p \,;\, n \leq p \leq N \wedge \tau_p(\omega) = p\}$$
$$= \inf\{p \,;\, n \leq p \leq N \wedge \mathbf{1}_{\{\tau_p = p\}} = 1\}(\omega).$$

2.4 Remark: Corresponding

Consider some $\tau \in \mathcal{T}$ and $C \in \mathcal{C}$ corresponding. Then we have $\tau \in \mathcal{S}(C)$.

2.5 Definition: Corresponding To

In the sequel the condition "corresponding" will be too strong several times. So we define a relaxation of that condition: For each $\tau \in \mathcal{T}$, $C \in \mathcal{C}$ and $n \in \mathbb{N}_0$ define

$$\tau_n \text{ corresponding to } C$$

iff
$$\tau_n = \inf \left\{ p \, ; \, n \leqslant p \leqslant N \wedge \mathbf{1}_{C_p} = 1 \right\}.$$

Consider some $\tau \in \mathcal{T}$ and $C \in \mathcal{C}$.
They are corresponding iff τ_n is corresponding to C for each $n \in \mathbb{N}_0$ with $n < N + 1$.

2.6 Definition and Remark: Essential

We will consider the general algorithm termination in Theorem 5.12 (page 61). We get better results if the initial parameters have a special type, which we will call essential. This is considered in Conclusion 5.14 (page 64). If the time horizon is finite and the initial parameters are of the most simple nature, there is again a small improvement. This is considered in Theorem 3.14 (page 39).

The special type of being "essential" mentioned above was first defined in [BKS08, part 3.1 ff.], therein only for elements of $\mathcal{Z}$ and called "a-priori-set". We will enlarge the definition to the corresponding elements of $\mathcal{C}$ and $\mathcal{T}$.

- $C \in \mathcal{C}$ is called essential if there is an optimal $\sigma \in \mathcal{T}$ such, that
  for all $n \in \mathbb{N}_0$ with $n < N + 1$ we have $\sigma_n \in \mathcal{S}(C)$.

- $\Theta \in \mathcal{Z}$ is called essential if there is an optimal $\sigma \in \mathcal{T}$ such, that
  for all $n \in \mathbb{N}_0$ with $n < N + 1$ $\sigma_n \in \Theta$ a.s., meaning $P(\{\omega \in \Omega \, ; \, \sigma_n(\omega) \in \Theta(\omega)\}) = 1$.

- $\tau \in \mathcal{T}$ is called essential if there is an optimal $\sigma \in \mathcal{T}$ such, that
  for all $n \in \mathbb{N}_0$ with $n < N + 1$ $\{\sigma_n = n\} \subseteq \{\tau_n = n\}$.

- Consider corresponding $C \in \mathcal{C}$, $\Theta \in \mathcal{Z}$ and $\tau \in \mathcal{T}$. Then each or none of them is essential.

- $\tau \in \mathcal{T}$ is essential if there is an optimal $\sigma \in \mathcal{T}$ with $\tau \leqslant \sigma$ point-wise (meaning $\tau_n \leqslant \sigma_n$ for all $n \in \mathbb{N}_0$ with $n < N + 1$). The following Lemma 2.7 (page 19) will show this.

2.7 Lemma: Elements of $\mathcal{T}$ compared

For all $\sigma, \tau \in \mathcal{T}$ we have the equivalence of the following assertions:

- $\{\sigma_n = n\} \subseteq \{\tau_n = n\}$ for all $n \in \mathbb{N}_0$ with $n < N + 1$,
- $\sigma_n \geq \tau_n$ for all $n \in \mathbb{N}_0$ with $n < N + 1$.

Proof:
Given the second assertion, the first follows instantaneously.

Consider the first assertion is true.
Let $n \in \mathbb{N}_0$ with $n < N + 1$.
We will prove $\sigma_n \geq \tau_n$ separately on $\{\tau_n = \infty\}$ and $\{\tau_n < \infty\}$.

First let $\omega \in \{\tau_n = \infty\}$ and assume $\omega \notin \{\sigma_n = \infty\}$. Due to Lemma 1.12 (page 13) there is $m \in \mathbb{N}_0$ with
$$\sigma_n(\omega) = \sigma_{n+m}(\omega) = n + m.$$
Since $\{\sigma_{n+m} = n + m\} \subseteq \{\tau_{n+m} = n + m\}$ it follows $\tau_{n+m}(\omega) = n + m$. Furthermore we have $\tau_{n+m}(\omega) \geq \tau_n(\omega)$ due to Lemma 1.12 (page 13). Hence $n + m = \sigma_{n+m}(\omega) \geq \tau_n(\omega) = \infty$, a contradiction. So we have $\sigma_n \geq \tau_n$ on $\{\tau_n = \infty\}$.

Due to Lemma 1.12 (page 13) we have
$$\{\tau_n < \infty\} = \bigcup_{m \in \mathbb{N}_0} \left(\{\tau_{n+m} = n + m\} \cap \bigcap_{k=n}^{n+m-1} \{\tau_k > k\} \right).$$

So it suffices to show by induction for all $m \in \mathbb{N}_0$
$$\sigma_n \geq \tau_n \quad \text{on} \quad \{\tau_{n+m} = n + m\} \cap \bigcap_{k=n}^{n+m-1} \{\tau_k > k\}.$$

For $m = 0$ the assertion is true, since $\{\sigma_n = n\} \subseteq \{\tau_n = n\}$ implies $\sigma_n \geq \tau_n$ on $\{\tau_n = n\}$.

Now consider it is true for some $m \in \mathbb{N}_0$. Then we have
$$\sigma_n = \sigma_{n+m+1} = \tau_{n+m+1} = \tau_n \quad \text{on} \quad \{\tau_{n+m+1} = n + m + 1\} \cap \{\tau_{n+m} > n + m\} \cap \bigcap_{k=n}^{n+m-1} \{\tau_k > k\}.$$

Hence it is true for $m + 1$. So we proved $\sigma_n \geq \tau_n$ on $\{\tau_n < \infty\}$. □

2.8 The Window Parameter

For each iteration of the algorithm step we fix some window parameter k. Let l be the minimum of k and the number of remaining time points ahead. We look l time points ahead in this manner: We compare the current pay-off with the expected pay-off when we do not use the next 1, 2, 3, ... l exercise possibilities and stop then according to the best stopping rule we have calculated until now. The window parameter k is the maximal number of time points we look ahead. It may be chosen differently in each step, so we therefore define a window parameter function

$$\kappa : \mathbb{N} \longrightarrow \mathbb{N} \cup \{\infty\}$$

as described at the beginning of this chapter, for example $\kappa \equiv 1$ or $\kappa \equiv N$.

Instead of fixing κ in advance, it is possible to choose $\kappa(m+1)$ dependent of the results of the first m iterations before doing the $(m+1)$th iteration. We do not examine this within this thesis because the "goodness" of the choice of κ depends strongly on the individual problem and the individual technique of calculating or approximating the involved conditional expectations.

2.9 Input for FII Algorithm

The FII Algorithm has this input parameters:

- the probability space $(\Omega, \mathcal{A}, P)$ with filtration $(\mathcal{A}_n)_{0 \leqslant n < \infty}$ described in Setting 1.2 (page 9),
- the process $(X_n)_{0 \leqslant n < \infty}$ described in Setting 1.2 (page 9),
- a window parameter function κ described in Part 2.8 (page 20).
- initial parameters $C^{(0)} \in \mathcal{C}$ and $\Theta^{(0)} \in \mathcal{Z}$ corresponding, and $\tau^{(0)} \in \mathcal{T}$.[1]

2.10 Remark: Input for FII Algorithm

The canonical initializing is done by setting

$$\tau_n^{(0)} \equiv n, C_n^{(0)} = \Omega \text{ and } \Theta_n^{(0)} \equiv \{i \; ; \; n \leqslant i \leqslant N\} \text{ for all } n \in \mathbb{N}_0 \text{ with } n < N+1$$
$$\text{and if } N = \infty \text{ also } C_\infty^{(0)} := \Omega.$$

When using any other $C^{(0)}$ the canonical definition for $\Theta^{(0)}$ and $\tau^{(0)}$ is choosing them corresponding by the mappings in Definition 2.3 (page 16).

---

[1] The idea of starting with some consistent family of stopping rules independently of choosing $C^{(0)}$ has been done first in [KS06, p. 34]. There it was done only for $C^{(0)} \equiv \Omega$ and finite N.

2.11 Output of FII Algorithm

In Part 2.13 (page 22) we describe how the algorithm is calculating the output. Here we show which form it has and in Remark 2.12 (page 21) we remark their meaning and sense.
The final output of the FII Algorithm is the tuple

$$\left(Y^{(\infty)}, \tau^{(\infty)}, \Theta^{(\infty)}, C^{(\infty)}\right)$$

with

$$Y^{(\infty)} = \left(Y_n^{(\infty)}\right)_{0 \leq n < N+1},$$
$$\tau^{(\infty)} = \left(\tau_n^{(\infty)}\right)_{0 \leq n < N+1},$$
$$\Theta^{(\infty)} = \left(\Theta_n^{(\infty)}\right)_{0 \leq n \leq N},$$
$$C^{(\infty)} = \left(C_n^{(\infty)}\right)_{0 \leq n \leq N}.$$

This is done by constructing iterative an approximation. In the mth iteration step ($m \in \mathbb{N}_0$) the interim output of the FII Algorithm is the tuple

$$\left(Y^{(m)}, \tilde{Y}^{(m)}, \hat{Y}^{(m)}, \tau^{(m)}, \hat{\tau}^{(m)}, \Theta^{(m)}, \hat{\Theta}^{(m)}, C^{(m)}, \hat{C}^{(m)}\right)$$

with $\tau^{(m)}, \hat{\tau}^{(m)}, Y^{(m)}, \tilde{Y}^{(m)}, \hat{Y}^{(m)}$ being of the form $\tau^{(m)} = \left(\tau_n^{(m)}\right)_{0 \leq n < N+1}$, and $\Theta^{(m)}, \hat{\Theta}^{(m)}, C^{(m)}, \hat{C}^{(m)}$ of the form $\Theta^{(m)} = \left(\Theta_n^{(m)}\right)_{0 \leq n \leq N}$.

2.12 Remarks: Output of FII Algorithm

- Given later discussed conditions for every $m \in \mathbb{N}$ the variable $Y^{(m)}$ is a lower approximation of the process Y^* as shown below.

- $\tau^{(m)}$ is a lower approximation of an (early) optimal stopping family. Under certain conditions it is even a lower approximation of $\tilde{\tau}^*$, the first optimal stopping family (see Theorem 3.14 (page 39)).

  $\tau^{(m)}$, $C^{(m)}$ and $\Theta^{(m)}$ are corresponding, hence each can be constructed from the other. They are just three different display forms of the same thing. Storing any of these three variables is enough to perform the next iteration step.

- $\hat{C}^{(m)}$, $\hat{\Theta}^{(m)}$, $\hat{\tau}^{(m)}$ are also corresponding and hence three different display forms of the same thing. In the finite case (for N finite) we will show in Theorem 3.8 (page 35) that each stopping rule between $\tau^{(m+1)}$ and $\hat{\tau}^{(m+1)}$ is an "improvement" of $\tau^{(m)}$. For getting an approximation of a stopping rule as small as possible we concentrate on the variables without hat.

- $\tilde{Y}^{(m)}$ and $\hat{Y}^{(m)}$ are auxiliary variables to construct the others.

2.13 Performing FII Algorithm

The calculation within the FII Algorithm is done this way:

- **Initializing Step** ($m = 0$): Set $\hat{\tau}^{(0)} := \tau^{(0)}$, $\hat{C}^{(0)} := C^{(0)}$, $\hat{\Theta}^{(0)} := \Theta^{(0)}$, and for all $n \in \mathbb{N}_0$ with $n < N + 1$ set

$$Y_n^{(0)} := \mathrm{E}\left(X_{\tau_n^{(0)}} \middle| \mathcal{A}_n\right) \text{ and } \tilde{Y}_n^{(0)} := \hat{Y}_n^{(0)} := 0.$$

- **Iteration Step** ($m > 0$): Suppose that for some $m \in \mathbb{N}_0$ the above mentioned tuple is constructed, then define for all $n \in \mathbb{N}_0$ with $n < N + 1$

$$\hat{Y}_n^{(m+1)} := \sup\left\{\mathrm{E}\left(X_{\tau_p^{(m)}} \middle| \mathcal{A}_n\right) \ ; \ n + 1 \leqslant p < \min\{n + \kappa(m), N\} + 1\right\}$$
$$= \sup\left\{\mathrm{E}\left(Y_p^{(m)} \middle| \mathcal{A}_n\right) \ ; \ n + 1 \leqslant p < \min\{n + \kappa(m), N\} + 1\right\},$$
$$\tilde{Y}_n^{(m+1)} := \sup\left\{\mathrm{E}\left(X_{\tau_p^{(m)}} \middle| \mathcal{A}_n\right) \ ; \ n \leqslant p < \min\{n + \kappa(m), N\} + 1\right\}$$
$$= \sup\left\{\mathrm{E}\left(Y_p^{(m)} \middle| \mathcal{A}_n\right) \ ; \ n \leqslant p < \min\{n + \kappa(m), N\} + 1\right\},$$

$$\hat{C}_n^{(m+1)} := \hat{C}_n^{(m)} \cap \left\{\hat{Y}_n^{(m+1)} < X_n\right\}$$
$$= \hat{C}_n^{(m)} \cap \bigcap\left\{\left\{\mathrm{E}\left(X_{\tau_p^{(m)}} \middle| \mathcal{A}_n\right) \leqslant X_n\right\} \ ; \ n + 1 \leqslant p < \min\{n + \kappa(m), N\} + 1\right\},$$
$$C_n^{(m+1)} := C_n^{(m)} \cap \left\{\tilde{Y}_n^{(m+1)} \leqslant X_n\right\}$$
$$= C_n^{(m)} \cap \bigcap\left\{\left\{\mathrm{E}\left(X_{\tau_p^{(m)}} \middle| \mathcal{A}_n\right) \leqslant X_n\right\} \ ; \ n \leqslant p < \min\{n + \kappa(m), N\} + 1\right\},$$

$$\Theta_n^{(m+1)} := \left\{p \ ; \ n \leqslant p \leqslant N \wedge \mathbf{1}_{C_p^{(m+1)}} = 1\right\} \qquad = \Theta_0^{(m+1)} \cap \{p \ ; \ n \leqslant p \leqslant N\},$$
$$\hat{\Theta}_n^{(m+1)} := \left\{p \ ; \ n \leqslant p \leqslant N \wedge \mathbf{1}_{\hat{C}_p^{(m+1)}} = 1\right\} \qquad = \hat{\Theta}_0^{(m+1)} \cap \{p \ ; \ n \leqslant p \leqslant N\},$$

$$\tau_n^{(m+1)} := \inf\left\{p \ ; \ n \leqslant p \leqslant N \wedge \mathbf{1}_{C_p^{(m+1)}} = 1\right\} \qquad = \inf \Theta_0^{(m+1)} \cap \{p \ ; \ n \leqslant p \leqslant N\}$$
$$\qquad\qquad\qquad\qquad\qquad\qquad\qquad\qquad\qquad = \inf \Theta_n^{(m+1)},$$
$$\hat{\tau}_n^{(m+1)} := \inf\left\{p \ ; \ n \leqslant p \leqslant N \wedge \mathbf{1}_{\hat{C}_p^{(m+1)}} = 1\right\} \qquad = \inf \hat{\Theta}_0^{(m+1)} \cap \{p \ ; \ n \leqslant p \leqslant N\}$$
$$\qquad\qquad\qquad\qquad\qquad\qquad\qquad\qquad\qquad = \inf \hat{\Theta}_n^{(m+1)},$$

$$Y_n^{(m+1)} := \mathrm{E}\left(X_{\tau_n^{(m+1)}} \middle| \mathcal{A}_n\right),$$

and if $N = \infty$, then define $\hat{C}_\infty^{(m+1)} := C_\infty^{(m+1)} := \Omega$ and $\Theta_\infty^{(m+1)} :\equiv \hat{\Theta}_\infty^{(m+1)} :\equiv \{\infty\}$ as well.

- **Final step** ($m = \infty$): Suppose that for all $m \in \mathbb{N}_0$ the above mentioned tuple is constructed. Define for all $n \in \mathbb{N}_0$ with $n < N+1$

$$C_n^{(\infty)} := \bigcap_{m \in \mathbb{N}_0} C_n^{(m)},$$

$$\Theta_n^{(\infty)} := \bigcap_{m \in \mathbb{N}_0} \Theta_n^{(m)},$$

$$\tau_n^{(\infty)} := \sup_{m \in \mathbb{N}} \tau_n^{(m)},$$

$$Y_n^{(\infty)} := \sup_{m \in \mathbb{N}_0} Y_n^{(m)}$$

and if $N = \infty$ define $C_\infty^{(\infty)} := \Omega$ and $\Theta_\infty^{(m+1)} :\equiv \{\infty\}$ as well.

2.14 Remarks: Performing FII Algorithm

- For all $m \in \mathbb{N}$ we have $C^{(m)} \in \mathcal{C}$, $\Theta^{(m)} \in \mathcal{Z}$, and $\tau^{(m)} \in \mathcal{T}$ and they are corresponding.

- For all $n \in \mathbb{N}_0$ with $n \leq N$ we have $C_n^{(\cdot)}$, $\Theta_n^{(\cdot)}$ non-increasing and $\tau_n^{(\cdot)}$ non-decreasing.

- For all $m \in \mathbb{N}$ we have $\hat{C}^{(m)} \in \mathcal{C}$, $\hat{\Theta}^{(m)} \in \mathcal{Z}$, and $\hat{\tau}^{(m)} \in \mathcal{T}$ and they are corresponding.

- For all $n \in \mathbb{N}_0$ with $n \leq N$ we have

$$\tau_n^{(\infty)} = \inf\left\{p \,;\, n \leq p < N+1 \wedge \mathbf{1}_{C_p^{(\infty)}} = 1\right\} \in \mathcal{S}(C^{(\infty)}).$$

2.15 Comparison with the Literature

There are some algorithms being similar to the one above suggested in the literature before. The variables therein examined in each step belong to those described in Part 2.11 (page 21).

The variables $\hat{Y}, \hat{C}, \hat{\Theta}$, and $\hat{\tau}$ only appear explicitly in [BS06, part 3.2 et seqq.]. The algorithm suggested in [Irl80, p. 180], [Irl06, pp. 102-103] and [Irl09] omits mentioning κ, Θ, Y and $\tilde{Y}$ explicitly. The algorithm suggested in [KS06, pp. 31-35,42] omits mentioning C explicitly, and a Θ very similar to ours appears therein in part six et seqq. The algorithm in [BKS08] omits mentioning κ and instead of examining Θ it just has a look at Θ_0, named A therein. Irle has a look at the infinite time horizon in his above mentioned papers, while the others focus on finite time horizon.

In the following lemma we collect elementary facts which we will use several times in the subsequent lemmas and theorems. Some of them are used implicitly here and there in [BS06, part 3.2, lemma 3.3].

2.16 Lemma: Elementary Equations and Inequalities

For all $n \in \mathbb{N}_0$ with $n < N+1$ and for all $m \in \mathbb{N}_0$ we have

$$\tilde{Y}_n^{(m+1)} = \max\left\{Y_n^{(m)}, \hat{Y}_n^{(m+1)}\right\}$$
$$= \max\left\{\mathrm{E}\left(X_{\tau_n^{(m)}}\middle|\mathcal{A}_n\right), \hat{Y}_n^{(m+1)}\right\}$$

(2.16.1) $$= \mathbf{1}_{\{\tau_n^{(m)}>n\}} \hat{Y}_n^{(m+1)} + \mathbf{1}_{\{\tau_n^{(m)}=n\}} \max\left\{\hat{Y}_n^{(m+1)}, X_n\right\}$$

(2.16.2) $$\geqslant \hat{Y}_n^{(m+1)},$$

(2.16.3) $$\left\{\tilde{Y}_n^{(m+1)} \leqslant X_n\right\} = \left\{\hat{Y}_n^{(m+1)} \leqslant X_n\right\},$$

$$C_n^{(m+1)} = C_n^{(m)} \cap \left\{\hat{Y}_n^{(m+1)} \leqslant X_n\right\}$$

(2.16.4) $$= C_n^{(m)} \cap \bigcap\left\{\left\{\mathrm{E}\left(X_{\tau_p^{(m)}}\middle|\mathcal{A}_n\right) \leqslant X_n\right\} ; n+1 \leqslant p < \min\{n+\kappa(m), N\}+1\right\},$$

(2.16.5) $$\mathrm{E}\left(\tilde{Y}_n^{(m)}\middle|\mathcal{A}_{n-1}\right) \geqslant \hat{Y}_{n-1}^{(m)}.$$

Proof:

- (2.16.1) follows due to the consistency of $\tau^{(m)}$ defined in Definition 1.10 (page 12),

$$\left\{\tau_n^{(m)} > n\right\} \subseteq \left\{\tau_n^{(m)} = \tau_{n+1}^{(m)}\right\} \text{ for all } n \in \mathbb{N}_0 \text{ with } n < N,$$

  by the definitions above and properties of conditional expectations.

- (2.16.3) follows by (2.16.1).
- (2.16.4) is just a combination of the underlying definitions and (2.16.3).
- For (2.16.5) have a look at the definitions. □

In Definition 2.18 (page 25) we will define "improver". To justify the definition we first state a lemma to show the existence of improvers. The idea of this definition comes from [BS06, definition 3.5].

2.17 Lemma: Existence of Improvers

For all $m \in \mathbb{N}_0$ and for all $n \in \mathbb{N}_0$ with $n < N+1$ we have
$$\hat{C}_n^{(m)} \subseteq C_n^{(m)},$$
$$\hat{\Theta}_n^{(m)} \subseteq \Theta_n^{(m)},$$
$$\tau_n^{(m)} \leq \hat{\tau}_n^{(m)}.$$

Proof: Assume $n \in \mathbb{N}_0$ with $n < N+1$.
Proof of the first statement by induction over m:
It is true for $m = 0$ by definition.
Suppose it is true for some $m \in \mathbb{N}_0$. We have
$$\left\{\hat{Y}_n^{(m+1)} < X_n\right\} \subseteq \left\{\hat{Y}_n^{(m+1)} \leq X_n\right\} \stackrel{2.16.3}{=} \left\{\tilde{Y}_n^{(m+1)} \leq X_n\right\}.$$

Hence it follows
$$C_n^{(m+1)} \stackrel{\text{def.}}{=} \left\{\tilde{Y}_n^{(m+1)} \leq X_n\right\} \cap C_n^{(m)} \supseteq \left\{\hat{Y}_n^{(m+1)} < X_n\right\} \cap \hat{C}_n^{(m)} \stackrel{\text{def.}}{=} \hat{C}_n^{(m+1)}.$$

The other statements are true since for all $m \in \mathbb{N}$ we have
$$\mathbf{1}_{\hat{C}_p^{(m+1)}} = 1 \implies \mathbf{1}_{C_p^{(m+1)}} = 1 \text{ for all } n \leq p \leq N,$$
hence
$$\Theta_n^{(m)} \stackrel{\text{def.}}{=} \left\{p \,;\, n \leq p \leq N \wedge \mathbf{1}_{C_p^{(m+1)}} = 1\right\} \supseteq \left\{p \,;\, n \leq p \leq N \wedge \mathbf{1}_{\hat{C}_p^{(m+1)}} = 1\right\} \stackrel{\text{def.}}{=} \hat{\Theta}_n^{(m)}$$
and
$$\tau_n^{(m)} \stackrel{\text{def.}}{=} \inf\left\{p \,;\, n \leq p \leq N \wedge \mathbf{1}_{C_p^{(m)}} = 1\right\} \leq \inf\left\{p \,;\, n \leq p \leq N \wedge \mathbf{1}_{\hat{C}_p^{(m)}} = 1\right\} \stackrel{\text{def.}}{=} \hat{\tau}_n^{(m)}. \quad \square$$

2.18 Definition: Improver

Consider $m \in \mathbb{N}$ and $\sigma \in \mathcal{T}$.
σ is called **improver of** $\tau^{(m)}$ iff
$$\tau_n^{(m+1)} \leq \sigma_n \leq \hat{\tau}_n^{(m+1)} \text{ for all } n \in \mathbb{N}_0 \text{ with } n < N+1.$$

Accordingly the term improver is defined for elements of $\mathcal{C}$ and $\mathcal{Z}$.

Remark: By Lemma 2.17 (page 25) $\tau^{(m+1)}$ and $\hat{\tau}^{(m+1)}$ are improvers of $\tau^{(m)}$.

2.19 Setting of [Irl80] included

If $N = \infty$ and $\kappa \equiv 1$, then for all $m \in \mathbb{N}_0$ and for all $n \in \mathbb{N}_0$ we have

$$C_n^{(m+1)} = \left\{ \mathrm{E}\left(X_{\tau_{n+1}^{(m)}} \Big| \mathcal{A}_n\right) \leq X_n \right\} \cap C_n^{(m)}.$$

Comment: Hence the algorithm of [Irl80] is included in the above described wider setting.

Proof: Consider $N = \infty$, $\kappa \equiv 1$, $m \in \mathbb{N}_0$ and $n \in \mathbb{N}_0$.
Then we have

$$\hat{Y}_n^{(m+1)} = \mathrm{E}\left(X_{\tau_{n+1}^{(m)}} \Big| \mathcal{A}_n\right),$$
$$\tilde{Y}_n^{(m+1)} = \sup \left\{ \mathrm{E}\left(X_{\tau_{n+1}^{(m)}} \Big| \mathcal{A}_n\right), \mathrm{E}\left(X_{\tau_n^{(m)}} \Big| \mathcal{A}_n\right) \right\}.$$

By Lemma 2.16 (page 24) we have

$$\left\{ \tilde{Y}_n^{(m+1)} \leq X_n \right\} = \left\{ \mathrm{E}\left(X_{\tau_{n+1}^{(m)}} \Big| \mathcal{A}_n\right) \leq X_n \right\},$$

hence

$$C_n^{(m+1)} = \left\{ \tilde{Y}_n^{(m+1)} \leq X_n \right\} \cap C_n^{(m)} = \left\{ \mathrm{E}\left(X_{\tau_{n+1}^{(m)}} \Big| \mathcal{A}_n\right) \leq X_n \right\} \cap C_n^{(m)}. \quad \square$$

2.20 Remark: Infinite Case Includes Finite Case

The case $N < \infty$ is of course part of the infinite case by setting $X_{N+n} := X_N$ for all $n \in \mathbb{N}$. Obviously it follows $X_\infty = X_N$.

3 Finite Time Horizon

In this chapter as well as in Section 5 (page 51) we examine further the behaviour of the output variables of our algorithm in dependence of the input variables. In particular we examine under which conditions the terminal output of FII algorithm is optimal in some manner. In this chapter we concentrate on the case of finite time horizon, while we will examine the general case in Section 5 (page 51).

In the first lemmas and theorems we state some fundamental results, we use later. In Remark 3.9 (page 37) and Theorem 3.10 (page 37) we show that $\tau^{(m)}$ is an increasing sequence which is (under weak conditions) lower than each optimal family of stopping times (of a certain class). Theorem 3.14 (page 39) and Theorem 3.18 (page 42) examine the algorithm termination. In Remark 3.15 (page 41) we have a look at an idea of [BKS08, Proposition 3.4] for making generalisations of these results, but we can show by Example 3.16 (page 41) and Example 3.17 (page 42), that this idea is not viable. In Theorem 3.21 (page 44) we show that with increasing κ the number of iterations decreases.

For the whole chapter consider finite time horizon, hence finite N. This finite case has been considered in [KS06] intensively, but always with $C^{(0)} \equiv \Omega$, since this parameter was not used explicitly. In this chapter we will remove this restriction. We will show that relaxing it to $\tau^{(0)}$ and $C^{(0)}$ corresponding is generally possible; for several assertions even this can often be relaxed to the assumption, that $\tau_n^{(0)}$ is corresponding to $C^{(0)}$ for several n.

In the following lemma we collect elementary facts we will use several times in the subsequent lemmas and theorems. Some of them are used implicitly here and there in [BS06, part 3.2] in the more simple form of $C^{(0)} \equiv \Omega$.

3.1 Theorem: Equations and Inequalities

We have for all $n \in \mathbb{N}_0$ with $n \leqslant N$

(3.1.1) $$X_n = Y_n^{(0)} \text{ on } \{\tau_n^{(0)} = n\},$$

(3.1.2) $$\tilde{Y}_n^{(1)} = \max\left\{\hat{Y}_n^{(1)}, X_n\right\} = \max\left\{\hat{Y}_n^{(1)}, Y_n^{(0)}\right\} \text{ on } \{\tau_n^{(0)} = n\},$$

(3.1.3) $$X_n < \tilde{Y}_n^{(1)} = \hat{Y}_n^{(1)} \text{ on } C_n^{(0)} \backslash C_n^{(1)},$$

(3.1.4) $$\hat{Y}_n^{(1)} = \tilde{Y}_n^{(1)} \text{ on } \{\tau_n^{(0)} > n\} \cup C_n^{(0)} \backslash C_n^{(1)},$$

(3.1.5) $$X_n = Y_n^{(0)} = \tilde{Y}_n^{(1)} \geqslant \hat{Y}_n^{(1)} \text{ on } \{\tau_n^{(0)} = n\} \cap C_n^{(1)}.$$

For all $m \in \mathbb{N}_0$ and $n \in \mathbb{N}_0$ with $n \leq N$ with $\tau_n^{(m)}$ corresponding to $C^{(m)}$ we have

(3.1.6) $$X_n = Y_n^{(m)} \text{ on } C_n^{(m)},$$
(3.1.7) $$\tilde{Y}_n^{(m+1)} = \max\left\{\hat{Y}_n^{(m+1)}, X_n\right\} = \max\left\{\hat{Y}_n^{(m+1)}, Y_n^{(m)}\right\} \text{ on } C_n^{(m)},$$
(3.1.8) $$X_n = Y_n^{(m)} < \tilde{Y}_n^{(m+1)} = \hat{Y}_n^{(m+1)} \text{ on } C_n^{(m)} \setminus C_n^{(m+1)},$$
(3.1.9) $$\tilde{Y}_n^{(m+1)} = \hat{Y}_n^{(m+1)} \text{ on } \left(C_n^{(m)}\right)^c,$$
(3.1.10) $$X_n = Y_n^{(m)} = \tilde{Y}_n^{(m+1)} \geq \hat{Y}_n^{(m+1)} \text{ on } C_n^{(m+1)}.$$

Proof: Consider $n \in \mathbb{N}_0$ with $n \leq N$, $m \in \mathbb{N}_0$. We will prove

(3.1.11) $$X_n = Y_n^{(m)} \text{ on } \{\tau_n^{(m)} = n\},$$
(3.1.12) $$\tilde{Y}_n^{(m+1)} = \max\left\{\hat{Y}_n^{(m+1)}, X_n\right\} \text{ on } \{\tau_n^{(m)} = n\},$$
(3.1.13) $$X_n < \tilde{Y}_n^{(m+1)} = \hat{Y}_n^{(m+1)} \text{ on } C_n^{(m)} \setminus C_n^{(m+1)},$$
(3.1.14) $$\tilde{Y}_n^{(m+1)} = \hat{Y}_n^{(m+1)} \text{ on } \{\tau_n^{(m)} > n\} \cup C_n^{(m)} \setminus C_n^{(m+1)},$$
(3.1.15) $$X_n = Y_n^{(m)} = \tilde{Y}_n^{(m+1)} \geq \hat{Y}_n^{(m+1)} \text{ on } \{\tau_n^{(m)} = n\} \cap C_n^{(m+1)}.$$

Before going into the details of the proof we shortly show the consequences:
The equations (3.1.1) to (3.1.5) are evident.
If $\tau_n^{(m)}$ is corresponding to $C^{(m)}$, we have

$$C_n^{(m+1)} \subset C_n^{(m)}, \ \{\tau_n^{(m)} = n\} = C_n^{(m)} \text{ and } \{\tau_n^{(m)} > n\} = \left(C_n^{(m)}\right)^c,$$

whereby equations (3.1.6), (3.1.7) and (3.1.8) instantly follow, and we have (3.1.9) due to

$$\{\tau_n^{(m)} > n\} \cup C_n^{(m)} \setminus C_n^{(m+1)} = \left(C_n^{(m)}\right)^c \cup C_n^{(m)} \setminus C_n^{(m+1)} = \left(C_n^{(m+1)}\right)^c$$

and (3.1.10) due to

$$\{\tau_n^{(m)} = n\} \cap C_n^{(m+1)} = C_n^{(m)} \cap C_n^{(m+1)} = C_n^{(m+1)}.$$

We have by the definitions in Algorithm 2.13 (page 22)

(3.1.16) $$\tau_n^{(m+1)} \stackrel{\text{def}}{=} \inf\left\{p \ ; \ n \leq p < N+1, \mathbf{1}_{C_p^{(m+1)}} = 1\right\},$$
(3.1.17) $$C_n^{(m+1)} \stackrel{\text{def}}{=} \left\{\tilde{Y}_n^{(m+1)} \leq X_n\right\} \cap C_n^{(m)},$$
(3.1.18) $$Y_n^{(m)} \stackrel{\text{def}}{=} \mathrm{E}\left(X_{\tau_n^{(m)}} \big| \mathcal{A}_n\right).$$

Proof of (3.1.11):
By properties of conditional expectations we have

$$\{\tau_n^{(m)} = n\} \subseteq \{Y_n^{(m)} = X_n\}.$$

Proof of (3.1.12):
This follows by Lemma 2.16 (page 24).

Proof of (3.1.13):
We have
$$\{\tau_n^{(m+1)} > n\} \cap C_n^{(m)}$$
$$= \left(C_n^{(m+1)}\right)^c \cap C_n^{(m)} \quad \text{by (3.1.16)}$$
$$= \left(\{\tilde{Y}_n^{(m+1)} \leqslant X_n\} \cap C_n^{(m)}\right)^c \cap C_n^{(m)} \quad \text{by (3.1.17)}$$
$$= \left(\{\tilde{Y}_n^{(m+1)} > X_n\} \cap C_n^{(m)}\right) \cup \left(\left(C_n^{(m)}\right)^c \cap C_n^{(m)}\right)$$
$$= \{\tilde{Y}_n^{(m+1)} > X_n\} \cap C_n^{(m)} \cup \emptyset$$
$$\subseteq \{\tilde{Y}_n^{(m+1)} > X_n\}.$$

So we have
$$X_n < \tilde{Y}_n^{(m+1)} \stackrel{2.16}{\leqslant} \max\left\{X_n, \hat{Y}_n^{(m+1)}\right\} \text{ on } \left(C_n^{(m+1)}\right)^c \cap C_n^{(m)}.$$
and thus
$$X_n < \tilde{Y}_n^{(m+1)} = \hat{Y}_n^{(m+1)} \text{ on } \left(C_n^{(m+1)}\right)^c \cap C_n^{(m)} = C_n^{(m)} \backslash C_n^{(m+1)}.$$

Proof of (3.1.14):
By Lemma 2.16 (page 24) we have
(3.1.19) $$\tilde{Y}_n^{(m+1)} = \hat{Y}_n^{(m+1)} \text{ on } \{\tau_n^{(m)} > n\}.$$
(3.1.14) is a combination of this fact with (3.1.13).

Proof of (3.1.15):
We have
$$\{\tau_n^{(m+1)} = n\} = C_n^{(m+1)} \quad \text{by (3.1.16)}$$
$$= \{\tilde{Y}_n^{(m+1)} \leqslant X_n\} \cap C_n^{(m)} \quad \text{by (3.1.17)}$$
(3.1.20) $$\stackrel{\text{triv.}}{\subseteq} \{\tilde{Y}_n^{(m+1)} \leqslant X_n\}.$$
By definition in Algorithm 2.13 (page 22) we have
(3.1.21) $$\tilde{Y}_n^{(m+1)} \geqslant Y_n^{(m)}.$$
So we have
$$X_n \stackrel{(3.1.20)}{\geqslant} \tilde{Y}_n^{(m+1)} \stackrel{(3.1.21)}{\geqslant} Y_n^{(m)} \stackrel{(3.1.11)}{=} X_n \text{ on } C_n^{(m+1)} \cap \{\tau_n^{(m)} = n\}.$$
Equality everywhere follows and thus
$$\max\left\{\hat{Y}_n^{(m+1)}, X_n\right\} \stackrel{(3.1.12)}{=} \tilde{Y}_n^{(m+1)} = Y_n^{(m)} = X_n \text{ on } C_n^{(m+1)} \cap \{\tau_n^{(m)} = n\}.$$
So we have
$$\hat{Y}_n^{(m+1)} \leqslant \tilde{Y}_n^{(m+1)} = Y_n^{(m)} = X_n \text{ on } C_n^{(m+1)} \cap \{\tau_n^{(m)} = n\}. \qquad \square$$

In the sequel we will show that our algorithm gives improved results in every step. We start with a look at the first iteration step and its specialties in the next lemma, Lemma 3.2 (page 30), and then proceed with the other steps in Theorem 3.3 (page 30) and therein also give an improvement for the first step under some conditions.

3.2 Lemma

For all $n \in \mathbb{N}_0$ with $n \leqslant N$ we have
$$\tilde{Y}_n^{(1)} \leqslant Y_n^{(1)} \quad \text{on} \quad \{\tau_n^{(0)} > n\} \cup C_n^{(0)}$$
and
$$X_n \leqslant Y_n^{(1)} \quad \text{on} \quad C_n^{(0)}.$$

Proof: Consider $n \in \mathbb{N}_0$ with $n \leqslant N$.
We have
$$\tilde{Y}_n^{(1)} \stackrel{\text{def.}}{\leqslant} X_n = Y_n^{(1)} \text{ on } C_n^{(1)} = \{\tau_n^{(1)} = n\},$$
$$X_n \stackrel{(3.1.3)}{<} \tilde{Y}_n^{(1)} \stackrel{(3.1.3)}{=} \hat{Y}_n^{(1)} \text{ on } C_n^{(0)} \setminus C_n^{(1)},$$
$$\hat{Y}_n^{(1)} \stackrel{(3.1.4)}{=} \tilde{Y}_n^{(1)} \text{ on } \{\tau_n^{(0)} > n\} \cup C_n^{(0)} \setminus C_n^{(1)}.$$

Consider some $n \in \mathbb{N}_0$ with $n < N - 1$ so that $\tilde{Y}_{n+1}^{(1)} \leqslant Y_{n+1}^{(1)}$. Then we have
$$\hat{Y}_n^{(1)} \stackrel{(3.3.2)}{\leqslant} Y_n^{(1)} \text{ on } \left(C_n^{(1)}\right)^c,$$
$$\left(\{\tau_n^{(0)} > n\} \cup C_n^{(0)} \setminus C_n^{(1)}\right) \cap \left(C_n^{(1)}\right)^c = \left(\{\tau_n^{(0)} > n\} \cup C_n^{(0)}\right) \cap \left(C_n^{(1)}\right)^c$$
and thus follows
$$\tilde{Y}_n^{(1)} \stackrel{3.1}{=} \hat{Y}_n^{(1)} \stackrel{(3.3.2)}{\leqslant} Y_n^{(1)} \text{ on } \left(\{\tau_n^{(0)} > n\} \cup C_n^{(0)}\right) \cap \left(C_n^{(1)}\right)^c = \{\tau_n^{(0)} > n\} \cup C_n^{(0)}. \qquad \square$$

Lemma 3.2 (page 30) above can be generalized: We state and prove here a theorem and a conclusion for arbitrary $C^{(0)}$, which has been proved for $C^{(0)} \equiv \Omega$ in [KS06, pp. 32-35]. We will generalize this theorem again in Theorem 3.8 (page 35) below.

3.3 Theorem

- For all $m \in \mathbb{N}$ and $n \in \mathbb{N}_0$ with $n \leqslant N$ we have
$$\tilde{Y}_n^{(m+1)} \leqslant Y_n^{(m+1)}.$$

- For all $n \in \mathbb{N}_0$ with $n \leqslant N$ and
$$\tau_k^{(0)} \text{ corresponding to } C^{(0)} \text{ for all } k \in \mathbb{N}_0 \text{ with } n \leqslant k \leqslant N$$
the inequality above is true for $m = 0$.

Proof via Backward Induction over n:
We will proof both points at once. So consider $m \in \mathbb{N}_0$.
We have
$$\tilde{Y}_N^{(m+1)} = \mathrm{E}\left(X_{\tau_N^{(m)}} \Big| \mathcal{A}_N\right) \quad \text{and} \quad Y_N^{(m+1)} = \mathrm{E}\left(X_{\tau_N^{(m+1)}} \Big| \mathcal{A}_N\right) \quad \text{and} \quad \tau_N^{(m)} \equiv N \equiv \tau_N^{(m+1)}.$$

Thus it follows
$$\tilde{Y}_N^{(m+1)} = \mathrm{E}\left(X_{\tau_N^{(m)}}\Big|\mathcal{A}_N\right) = \mathrm{E}\left(X_N|\mathcal{A}_N\right) = X_N = \mathrm{E}\left(X_N|\mathcal{A}_N\right) = \mathrm{E}\left(X_{\tau_N^{(m+1)}}\Big|\mathcal{A}_N\right) = Y_N^{(m+1)}.$$

Consider some $n \in \mathbb{N}_0$ with $n < N-1$ so that $\tilde{Y}_{n+1}^{(m+1)} \leqslant Y_{n+1}^{(m+1)}$.

We have
$$\tilde{Y}_n^{(m+1)} \stackrel{\text{def.}}{\leqslant} X_n = Y_n^{(m+1)} \text{ on } C_n^{(m+1)} = \{\tau_n^{(m+1)} = n\}.$$

Due to

(3.3.1)
$$\left(C_n^{(m+1)}\right)^c = \{\tau_n^{(m+1)} > n\} \subseteq \{\tau_n^{(m+1)} = \tau_{n+1}^{(m+1)}\},$$

it follows on $\left(C_n^{(m+1)}\right)^c$:

$$\begin{aligned}
Y_n^{(m+1)} &= \mathrm{E}\left(X_{\tau_n^{(m+1)}}\Big|\mathcal{A}_n\right) && \text{by def.} \\
&= \mathrm{E}\left(X_{\tau_{n+1}^{(m+1)}}\Big|\mathcal{A}_n\right) && \text{by (3.3.1)} \\
&= \mathrm{E}\left(\mathrm{E}\left(X_{\tau_{n+1}^{(m+1)}}\Big|\mathcal{A}_{n+1}\right)\Big|\mathcal{A}_n\right) && \\
&= \mathrm{E}\left(Y_{n+1}^{(m+1)}\Big|\mathcal{A}_n\right) && \text{by def.} \\
&\geqslant \mathrm{E}\left(\tilde{Y}_{n+1}^{(m+1)}\Big|\mathcal{A}_n\right) && \text{by ind. ass.} \\
&\geqslant \hat{Y}_n^{(m+1)} && \text{by 2.16.}
\end{aligned}$$

(3.3.2)

If $m > 0$ or $m = 0$ and $\tau_n^{(0)}$ is corresponding to $C^{(0)}$ we have

$$\tilde{Y}_n^{(m+1)} \stackrel{(3.1.9)}{=} \hat{Y}_n^{(m+1)} \stackrel{(3.3.2)}{\leqslant} Y_n^{(m+1)} \text{ on } \left(C_n^{(m+1)}\right)^c. \qquad \square$$

The following conclusion has been proved in [KS06, pp. 32-35] for $C^{(0)} \equiv \Omega$. We will expand it to the case of general $C^{(0)}$.

3.4 Conclusion

For all $m \in \mathbb{N}$ and $n \in \mathbb{N}_0$ with $n \leqslant N$ we have

$$Y_n^{(m)} \leqslant \tilde{Y}_n^{(m+1)} \leqslant Y_n^{(m+1)} \leqslant Y_n^*,$$

$$\tilde{Y}_n^{(m+1)} \leqslant \tilde{Y}_n^{(m+2)},$$

$$\hat{Y}_n^{(m+1)} \leqslant \hat{Y}_n^{(m+2)}.$$

For all $n \in \mathbb{N}_0$ with $n \leqslant N$ and

$$\tau_k^{(0)} \text{ corresponding to } C^{(0)} \text{ for all } k \in \mathbb{N}_0 \text{ with } n \leqslant k \leqslant N$$

the inequalities above are true for $m = 0$.

Proof: Consider $n \in \mathbb{N}_0$ with $n \leqslant N$ and $m \in \mathbb{N}_0$. We have by definition
$$Y_n^{(m)} \leqslant \tilde{Y}_n^{(m+1)} \text{ and } Y_n^{(m+1)} \leqslant Y_n^*.$$
The proof of
$$\tilde{Y}_n^{(m+1)} \leqslant Y_n^{(m+1)}$$
is done for $m > 0$ as well as for $m = 0$ under the mentioned condition in Theorem 3.3 (page 30).

The next inequality follows by repeating the arguments of the first set of inequalities.

By the first set of inequalities we have $Y_p^{(m)} \leqslant Y_p^{(m+1)}$ for all $p \in \mathbb{N}_0$ with $n+1 \leqslant p \leqslant N$. Hence the last line follows by the definitions of $\hat{Y}$ in Algorithm 2.13 (page 22). $\square$

In the following example we will show that Lemma 3.2 (page 30) cannot be expanded to a bigger subset of Ω, and that the condition of correspondency in Theorem 3.3 (page 30) and in Formula (3.1.6) (page 28) cannot be dropped. It will also be helpful for checking that conditions given in the following theorems are sharp.

3.5 Example

The following example will show the possibility of
$$Y_n^{(1)} = \hat{Y}_n^{(1)} < X_n = \tilde{Y}_n^{(1)} \text{ on } \{\tau_n^{(0)} = n\} \cap \left(C_n^{(0)}\right)^c = \{\tau_n^{(0)} = n\} \cap \left(C_n^{(1)}\right)^c.$$

Example: Consider $N \in \mathbb{N}_0 \cup \{\infty\}$ and $n \in \mathbb{N}_0$ with $n < N$ and a setting with
$$X_n, X_{n+1} \text{ deterministic with } X_n > X_{n+1},$$
$$\kappa(0) = 1,$$
$$\tau_{n+1}^{(0)} \equiv n+1,$$
$$C_n^{(0)} \subsetneq \{\tau_n^{(0)} = n\},$$
then we have
$$\hat{Y}_n^{(1)} \stackrel{\text{def}}{=} \sup\left\{\mathrm{E}\left(X_{\tau_p^{(0)}}\middle|\mathcal{A}_n\right) \;;\; n+1 \leqslant p < \min\{n+\kappa(m), N\} + 1\right\}$$
$$= \sup\left\{\mathrm{E}\left(X_{\tau_p^{(0)}}\middle|\mathcal{A}_n\right) \;;\; n+1 \leqslant p < n+2\right\}$$
$$= \mathrm{E}\left(X_{\tau_{n+1}^{(0)}}\middle|\mathcal{A}_n\right)$$
$$\stackrel{\text{def}}{=} \mathrm{E}\left(X_{n+1}\middle|\mathcal{A}_n\right)$$
$$= X_{n+1}$$
$$< X_n$$
and it follows
$$\hat{Y}_n^{(1)} < X_n = \max\left\{X_n, \hat{Y}_n^{(1)}\right\} = \tilde{Y}_n^{(1)} \text{ on } \left(\{\tau_n^{(0)} > n\} \cup C_n^{(0)}\right)^c = \{\tau_n^{(0)} = n\} \cap \left(C_n^{(0)}\right)^c \neq \emptyset.$$

Assume additionally
$$n+1 < N,$$
$$X_{n+2} = X_{n+1},$$
$$\tau^{(0)}_{n+2} \equiv n+2,$$
$$C^{(0)}_n \subsetneq C^{(0)}_{n+1} = \Omega.$$

Then we have
$$\hat{Y}^{(1)}_{n+1} = X_{n+2} = X_{n+1}$$

and
$$\begin{aligned}
\tilde{Y}^{(1)}_n &\stackrel{\text{def}}{=} \sup\left\{ \mathrm{E}\left(X_{\tau^{(0)}_p}\big|\mathcal{A}_n\right) \,;\, n \leqslant p < \min\{n+\kappa(0), N\}+1\right\} \\
&\stackrel{\text{def}}{=} \max\left\{ \mathrm{E}\left(X_{\tau^{(0)}_p}\big|\mathcal{A}_n\right) \,;\, n \leqslant p < n+2\right\} \\
&= \max\left\{ \mathrm{E}\left(X_{\tau^{(0)}_n}\big|\mathcal{A}_n\right), \mathrm{E}\left(X_{\tau^{(0)}_{n+1}}\big|\mathcal{A}_n\right) \right\} \\
&= \max\left\{ X_n \mathbf{1}_{\{\tau^{(0)}_n = n\}} + X_{n+1} \mathbf{1}_{\{\tau^{(0)}_n = n+1\}}, X_{n+1} \right\} \\
&= X_n \mathbf{1}_{\{\tau^{(0)}_n = n\}} + X_{n+1} \mathbf{1}_{\{\tau^{(0)}_n = n+1\}},
\end{aligned}$$

thus
$$\tilde{Y}^{(1)}_n = X_n \text{ on } \{\tau^{(0)}_n = n\}.$$

We have
$$C^{(1)}_n = \left\{ \tilde{Y}^{(1)}_n \leqslant X_n \right\} \cap C^{(0)}_n = \Omega \cap C^{(0)}_n = C^{(0)}_n$$

and thus
$$\begin{aligned}
C^{(1)}_{n+1} &= \left\{ \tilde{Y}^{(1)}_{n+1} \leqslant X_n \right\} \cap C^{(0)}_{n+1} \\
&= \left\{ \tilde{Y}^{(1)}_{n+1} \leqslant X_n \right\} \\
&\stackrel{2.16}{=} \left\{ \hat{Y}^{(1)}_{n+1} \leqslant X_n \right\} \\
&= \{X_{n+1} \leqslant X_n\} \\
&= \Omega
\end{aligned}$$

and
$$\{\tau^{(1)}_n = n+1\} = C^{(1)}_{n+1} \backslash C^{(1)}_n = \left(C^{(1)}_n\right)^c = \left(C^{(0)}_n\right)^c.$$

Due to
$$X_{n+1} = Y^{(1)}_n \text{ on } \{\tau^{(1)}_n = n+1\}$$

it follows that
$$\tilde{Y}^{(1)}_n = X_n > X_{n+1} = Y^{(1)}_n \text{ on } \{\tau^{(0)}_n = n\} \cap \left(C^{(0)}_n\right)^c. \qquad \square$$

3.6 Lemma

Fix some $n \in \mathbb{N}_0$.
For all $p \in \mathbb{N}$ and $m \in \mathbb{N}_0$ with $p \leq m$ we have

$$C_n^{(m+1)} = \left\{ \tilde{Y}_n^{(m+1)} \leq X_n \right\} \cap C_n^{(p)},$$

$$\hat{C}_n^{(m+1)} = \left\{ \hat{Y}_n^{(m+1)} < X_n \right\} \cap \hat{C}_n^{(p)}.$$

If
$$\tau_k^{(0)} \text{ is corresponding to } C^{(0)} \text{ for all } k \in \mathbb{N}_0 \text{ with } n \leq k \leq N,$$
we have for all $m \in \mathbb{N}_0$

$$C_n^{(m+1)} = \left\{ \tilde{Y}_n^{(m+1)} \leq X_n \right\} \cap C_n^{(0)},$$

$$\hat{C}_n^{(m+1)} = \left\{ \hat{Y}_n^{(m+1)} < X_n \right\} \cap C_n^{(0)}.$$

Proof by Induction over m:
Fix some $p \in \mathbb{N}_0$.
The hypothesis is trivially true for $m = p$ by definition.
Assume it is true for some $m \in \mathbb{N}_0$ with $p \leq m$. We have by Conclusion 3.4 (page 31)

$$\tilde{Y}_n^{(m+1)} \leq \tilde{Y}_n^{(m+2)},$$

$$\hat{Y}_n^{(m+1)} \leq \hat{Y}_n^{(m+2)}.$$

Thus

$$\left\{ \tilde{Y}_n^{(m+2)} \leq X_n \right\} \subseteq \left\{ \tilde{Y}_n^{(m+1)} \leq X_n \right\},$$

$$\left\{ \hat{Y}_n^{(m+2)} < X_n \right\} \subseteq \left\{ \hat{Y}_n^{(m+1)} < X_n \right\}.$$

So we have

$$C_n^{(m+2)} \stackrel{\text{def.}}{=} \left\{ \tilde{Y}_n^{(m+2)} \leq X_n \right\} \cap C_n^{(m+1)}$$
$$= \left\{ \tilde{Y}_n^{(m+2)} \leq X_n \right\} \cap \left\{ \tilde{Y}_n^{(m+1)} \leq X_n \right\} \cap C_n^{(p)}$$
$$= \left\{ \tilde{Y}_n^{(m+2)} \leq X_n \right\} \cap C_n^{(p)}$$

and

$$\hat{C}_n^{(m+2)} \stackrel{\text{def.}}{=} \left\{ \hat{Y}_n^{(m+2)} < X_n \right\} \cap \hat{C}_n^{(m+1)}$$
$$= \left\{ \hat{Y}_n^{(m+2)} < X_n \right\} \cap \left\{ \hat{Y}_n^{(m+1)} < X_n \right\} \cap \hat{C}_n^{(p)}$$
$$= \left\{ \hat{Y}_n^{(m+2)} < X_n \right\} \cap \hat{C}_n^{(p)}.$$

Due to Algorithm 2.13 (page 22) we have $\hat{C}_n^{(0)} = C_n^{(0)}$. □

The following theorem has been proved for the special case of
$$C^{(0)} \equiv \Omega \quad \text{and} \quad \tau^{(0)} \text{ and } C^{(0)} \text{ corresponding}$$
in [KS06, pp. 32-35]. Here we consider arbitrary $C^{(0)}$ and $\tau^{(0)}$.

3.7 Theorem

For all $n \in \mathbb{N}_0$ with $n \leq N$ and $m \in \mathbb{N}$ we have
$$X_n \leq \tilde{Y}_n^{(m+1)} \text{ on } C_n^{(0)}.$$
For all $n \in \mathbb{N}_0$ with $n \leq N$ and
$$\tau_k^{(0)} \text{ corresponding to } C^{(0)} \text{ for all } k \in \mathbb{N}_0 \text{ with } n \leq k \leq N$$
the set of inequalities above is true for $m = 0$.

Proof: Let $m \in \mathbb{N}$. Then we have
$$X_n \stackrel{(3.1.6)}{=} Y_n^{(1)} \text{ on } C_n^{(1)},$$
$$X_n \stackrel{(3.1.3)}{<} \tilde{Y}_n^{(1)} \text{ on } C_n^{(0)} \setminus C_n^{(1)},$$
$$Y_n^{(1)} \stackrel{3.4}{\leq} \tilde{Y}_n^{(2)},$$
$$\tilde{Y}_n^{(1)} \stackrel{3.4}{\leq} \tilde{Y}_n^{(2)};$$
hence
$$X_n \leq \tilde{Y}_n^{(2)} \stackrel{3.3}{\leq} Y_n^{(2)} \stackrel{3.4}{\leq} \tilde{Y}_n^{(m+1)} \text{ for all } m \in \mathbb{N}.$$
Under the mentioned additional assumption we have
$$X_n \stackrel{(3.1.12)}{\leq} \tilde{Y}_n^{(1)} \text{ on } \{\tau_n^{(0)} = n\} = C_n^{(0)}. \qquad \square$$

In Theorem 3.3 (page 30) and Theorem 3.7 (page 35) we had a look at $\tau^{(m+1)}$, a special improver of $\tau^{(m)}$ (for each $m \in \mathbb{N}_0$). Now we will have a look at an arbitrary improver and state the same result. This is based on an idea in [BS06, Part 3.2, Theorem 3.4 and Remark 3.4].

3.8 Theorem

Consider $m \in \mathbb{N}_0$ and $\sigma \in \mathcal{T}$ being an improver of $\tau^{(m)}$.
For all $n \in \mathbb{N}_0$ with $n \leq N$ and
$$\tau_k^{(0)} \text{ corresponding to } C^{(0)} \text{ for all } k \in \mathbb{N}_0 \text{ with } n \leq k \leq N$$
we have
$$\tilde{Y}_n^{(m+1)} \leq \mathrm{E}\left(X_{\sigma_n} | \mathcal{A}_n\right)$$
and
$$X_n \leq \mathrm{E}\left(X_{\sigma_n} | \mathcal{A}_n\right) \text{ on } C_n^{(0)}.$$

Proof of the first statement via backward induction over n:
We have by definition
$$\tilde{Y}_N^{(m+1)} = \mathrm{E}\left(X_{\tau_N^{(m+1)}}\middle|\mathcal{A}_N\right),$$
$$\tau_N^{(m+1)} \equiv N \equiv \sigma_N.$$
So we have
$$\tilde{Y}_N^{(m+1)} = X_N = \mathrm{E}\left(X_{\sigma_N}\middle|\mathcal{A}_N\right).$$
Suppose $n \in \mathbb{N}_0$ with $n < N$ such that
$$\tilde{Y}_{n+1}^{(m+1)} \leq \mathrm{E}\left(X_{\sigma_{n+1}}\middle|\mathcal{A}_{n+1}\right).$$
Due to $n \leq \tau_n^{(m+1)} \leq \sigma_n$ we have
$$\{\sigma_n = n\} \subseteq \{\tau_n^{(m+1)} = n\} = C_n^{(m+1)}$$
and by properties of conditional expectations we have
$$\{\sigma_n = n\} \subseteq \{\mathrm{E}(X_{\sigma_n}|\mathcal{A}_n) = X_n\}.$$
Thus it follows
$$\tilde{Y}_n^{(m+1)} \leq \mathrm{E}\left(X_{\sigma_n}\middle|\mathcal{A}_n\right) = X_n \text{ on } \{\sigma_n = n\}.$$
We have $\{\sigma_n > n\} \subseteq \{\sigma_n = \sigma_{n+1}\}$, hence on $\{\sigma_n > n\}$
$$\mathrm{E}\left(X_{\sigma_n}\middle|\mathcal{A}_n\right) = \mathrm{E}\left(X_{\sigma_{n+1}}\middle|\mathcal{A}_n\right) = \mathrm{E}\left(\mathrm{E}\left(X_{\sigma_{n+1}}\middle|\mathcal{A}_{n+1}\right)\middle|\mathcal{A}_n\right)$$
$$\geq \mathrm{E}\left(\tilde{Y}_{n+1}^{(m+1)}\middle|\mathcal{A}_n\right) \qquad \text{by ind.hyp.}$$
$$\geq \hat{Y}_n^{(m+1)} \qquad \text{by 2.16.}$$
On $\{\tau_n^{(m)} > n\}$ we have $\hat{Y}_n^{(m+1)} = \tilde{Y}_n^{(m+1)}$.
Hence we have $\mathrm{E}\left(X_{\sigma_n}\middle|\mathcal{A}_n\right) \geq \hat{Y}_n^{(m+1)} = \tilde{Y}_n^{(m+1)}$ on $\{\sigma_n > n\} \cap \{\tau_n^{(m)} > n\}$.
Furthermore we have
$$\{\sigma_n > n\} \subseteq \{\hat{\tau}_n^{(m+1)} > n\} = \left(\hat{C}_n^{(m+1)}\right)^c \overset{3.6}{=} \left(\{\hat{Y}_n^{(m+1)} < X_n\} \cap C_n^{(0)}\right)^c$$
$$= \{\hat{Y}_n^{(m+1)} \geq X_n\} \cup \left(C_n^{(0)}\right)^c.$$
Since $m > 0$ or $\tau_n^{(0)}$ is corresponding to $C^{(0)}$ we have
$$\{\tau_n^{(m)} = n\} = C_n^{(m)} \subseteq C_n^{(0)}$$
and thus
$$\{\sigma_n > n\} \cap \{\tau_n^{(m)} = n\} \subseteq \left(\{\hat{Y}_n^{(m+1)} \geq X_n\} \cup \left(C_n^{(0)}\right)^c\right) \cap C_n^{(0)}$$
$$\subseteq \{\hat{Y}_n^{(m+1)} \geq X_n\}.$$
So we have
$$\mathrm{E}\left(X_{\sigma_n}\middle|\mathcal{A}_n\right) \geq \hat{Y}_n^{(m+1)} \geq X_n \text{ on } \{\sigma_n > n\} \cap \{\tau_n^{(m)} = n\}$$
and hence
$$\mathrm{E}\left(X_{\sigma_n}\middle|\mathcal{A}_n\right) \geq \max\left\{X_n, \hat{Y}_n^{(m+1)}\right\} \overset{(3.1.12)}{=} \tilde{Y}_n^{(m+1)} \text{ on } \{\sigma_n > n\} \cap \{\tau_n^{(m)} = n\}. \qquad \square$$

Proof of the second statement:
We have for all $n \in \mathbb{N}_0$ with $n \leqslant N$

$$X_n \overset{3.7}{\leqslant} \tilde{Y}_n^{(m+1)} \leqslant \mathrm{E}\left(X_{\sigma_n}|\mathcal{A}_n\right) \text{ on } C_n^{(0)}.$$ □

3.9 Remark

Consider $n \in \mathbb{N}_0$ with $n \leqslant N$ and $m \in \mathbb{N}$.
Then we have
$$\tau_n^{(m)} \leqslant \tau_n^{(m+1)} \leqslant \tau_n^{(\infty)}.$$
If
$$\tau_k^{(0)} \text{ is corresponding to } C^{(0)} \text{ for all } k \in \mathbb{N}_0 \text{ with } n \leqslant k \leqslant N$$
this is true for $m = 0$ as well.

The next theorem shows that under certain conditions the limiting value of this increasing sequence of stopping rules is smaller than certain optimal stopping rules. We shall see later (in Theorem 3.18 (page 42)) that this limit is equal to the smalles optimal stopping rule. This theorem was proved for $C^{(0)} \equiv \Omega$ in [KS06, p. 35; Prop. 4.1], here we state it for a more arbitrary situation.

3.10 Theorem

Consider $n \in \mathbb{N}_0$ with $n \leqslant N$ and
$$\tau_k^{(0)} \text{ corresponding to } C^{(0)} \text{ for all } k \in \mathbb{N}_0 \text{ with } n \leqslant k \leqslant N$$
and some optimal $\sigma \in \mathcal{S}_n^N \cap \mathcal{S}(C^{(0)})$.
Then we have
$$\tau_n^{(\infty)} \leqslant \sigma.$$

Proof: We have
$$\tau_n^{(m)} = \inf\left\{p\,;\, n \leqslant p \leqslant N \wedge \mathbf{1}_{C_p^{(m)}} = 1\right\}$$
and due to Lemma 3.6 (page 34)
$$C_n^{(m)} = \left\{\tilde{Y}_n^{(m)} \leqslant X_n\right\} \cap C_n^{(0)},$$
thus we have
$$\tau_n^{(m)} = \inf\left\{p\,;\, n \leqslant p \leqslant N \wedge \tilde{Y}_p^{(m)} \leqslant X_p \wedge \mathbf{1}_{C_p^{(0)}} = 1\right\}.$$
Combined with the assumption that $\sigma \in \mathcal{S}(C^{(0)})$, we have due to the infimum above
$$\left\{\tilde{Y}_\sigma^{(m)} \leqslant X_\sigma\right\} \subseteq \left\{\tau_n^{(m)} \leqslant \sigma\right\}.$$
We have
$$X_\sigma = Y_\sigma^*$$

due to the optimality of σ and
$$Y_\sigma^* \geq \tilde{Y}_\sigma^{(m)}$$
due to the definiton of Y^* in Definition 1.3 (page 9) and the definition of $\tilde{Y}^{(m)}$ in Algorithm 2.13 (page 22). Hence
$$\tilde{Y}_\sigma^{(m)} \leq X_\sigma,$$
which yields $\tau_n^{(m)} \leq \sigma$. This is true for all $m \in \mathbb{N}$, thus we have $\tau_n^{(\infty)} \leq \sigma$ by definition. $\square$

The next lemma has been proved for the special case of $C^{(0)} \equiv \Omega$ and $\sigma = \tau^{(m+1)}$ via backward induction in [KS06, p. 36; lemma 4.2] which is generalized here to an arbitrary improver σ.

3.11 Lemma

Consider $m \in \mathbb{N}_0$, $n \in \mathbb{N}_0$ with $n \leq N$ and

$\tau_k^{(0)}$ corresponding to $C^{(0)}$ for all $k \in \mathbb{N}_0$ with $n \leq k \leq N$.

We have for all $\sigma \in \mathcal{T}$ being an improver of $\tau^{(m)}$

$$\mathrm{E}\left(X_{\sigma_n}|\mathcal{A}_n\right) \geq \tilde{Y}_n^{(m+1)} \geq \mathrm{E}\left(Y_{n+1}^{(m)}\Big|\mathcal{A}_n\right),$$

thus it follows

$$Y_n^{(m+1)} \geq \tilde{Y}_n^{(m+1)} \geq \mathrm{E}\left(Y_{n+1}^{(m)}\Big|\mathcal{A}_n\right).$$

Proof: The inequality on the left side is true due to Theorem 3.8 (page 35) and Theorem 3.3 (page 30) respectively, the other one can easily be verified by a look at the definitions. $\square$

3.12 Lemma

Let $\sigma \in \left(\mathcal{S}_0^N\right)^{\mathbb{N}_0}$ non-decreasing. Then we have

$$\lim_{n \to \infty} X_{\sigma_n} = X_{\lim_{n \to \infty} \sigma_n}$$

and

$$\mathrm{E}\left(X_{\lim_{n \to \infty} \sigma_n}\Big|\mathcal{A}_j\right) = \lim_{n \to \infty} \mathrm{E}\left(X_{\sigma_n}|\mathcal{A}_j\right) \text{ for all } j \in \mathbb{N}_0.$$

Proof: For all $n \in \mathbb{N}_0$ σ_n is valued in the discrete set $\{0, ..., N\}$. Hence obviously

$$\{\sup\{\sigma_n \ ; \ n \in \mathbb{N}_0\} \leq N < \infty\} = \Omega,$$

which implies the first line. By Lemma 1.4 (page 10) $\sup_{n \in \mathbb{N}_0} |X_{\sigma_n}|$ is integrable. So dominated convergence for conditional expectations shows the second assertion. (For details of this argument see [Irl05, pp. 126, 127].) $\square$

Now we want to generalize a statement made in [KS06, p. 36] for $C^{(0)} \equiv \Omega$. This condition can be dropped without any change to the statement itself.

3.13 Conclusion

For all $n \in \mathbb{N}_0$ with $n \leqslant N$ we have

$$Y_n^{(\infty)} = \mathrm{E}\left(X_{\tau_n^{(\infty)}} \Big| \mathcal{A}_n\right).$$

Proof: This follows directly by Lemma 3.12 (page 38) with $\sigma = (\tau_n^{(m)})_{m \in \mathbb{N}_0}$. $\square$

Next we show that the algorithm terminates with optimal results when started with $C^{(0)} \equiv \Omega$. A condensed proof can be found in [KS06, pp. 36-37; Proposition 4.3 and Proposition 4.4]; we give herein an extensive proof.

3.14 Theorem: Optimal Algorithm Termination for $C^{(0)} \equiv \Omega$

Consider $C^{(0)} \equiv \Omega$ and $\tau^{(0)}$ and $C^{(0)}$ corresponding.
For all $n \in \mathbb{N}_0$ with $n \leqslant N$ and $m \in \mathbb{N}_0$ with $N - n \leqslant m$ we have

$$\tau_n^{(m)} = \tau_n^{(\infty)} = \check{\tau}_n^*,$$

$$Y_n^{(m)} = Y_n^{(\infty)} = Y_n^*.$$

Proof: We will prove both sets of equations by backward induction over n.

The latter set of equations:
We have

$$\tau_N^{(0)} = \inf\left\{p\,;\, N \leqslant p \leqslant N, \mathbb{1}_{C_p^{(0)}} = 1\right\} \equiv N$$

and hence for all $m \in \mathbb{N}_0$ with $0 = N - N \leqslant m$

$$Y_N^{(m)} = Y_N^{(\infty)} = Y_N^* = X_N.$$

Consider some $n \in \mathbb{N}_0$ with $n < N$ and assume that the assertion is proved for $n + 1$.
Then we have $1 \leqslant N - n$.
Consider $m \in \mathbb{N}_0$ with $N - n \leqslant m$, hence we have $1 \leqslant m$.
Due to $N - (n+1) \leqslant m - 1$ and $0 \leqslant m - 1$ we have

(3.14.1) $$Y_{n+1}^{(m-1)} = Y_{n+1}^*$$

by induction hypothesis.

Consider a consistent N-suitable family of stopping rules σ, being an improver of $\tau^{(m-1)}$.
Since $\tau^{(0)}$ and $C^{(0)}$ are corresponding it follows

$$\mathrm{E}\left(X_{\sigma_n}|\mathcal{A}_n\right) \overset{3.11}{\geqslant} \mathrm{E}\left(Y_{n+1}^{(m-1)}\Big|\mathcal{A}_n\right) \overset{(3.14.1)}{=} \mathrm{E}\left(Y_{n+1}^*\Big|\mathcal{A}_n\right)$$

and due to $C_n^{(0)} \equiv \Omega$
$$\mathrm{E}\left(X_{\sigma_n}|\mathcal{A}_n\right) \stackrel{3.8}{\geqslant} X_n.$$
Thus it follows
$$Y_n^* \geqslant \mathrm{E}\left(X_{\sigma_n}|\mathcal{A}_n\right) \geqslant \max\left\{X_n, \mathrm{E}\left(Y_{n+1}^*\big|\mathcal{A}_n\right)\right\} \stackrel{1.7}{=} Y_n^*,$$
hence
$$Y_n^* = \mathrm{E}\left(X_{\sigma_n}|\mathcal{A}_n\right).$$
This is especially true for $\sigma = \tau^{(m)}$, hence we have $Y_n^* = Y_n^{(m)}$.

So $Y_n^* = Y_n^{(m)}$ for all $m \in \mathbb{N}_0$ with $N - n \leqslant m$ and $1 < m$, which implies
$$Y_n^{(\infty)} = Y_n^*.$$

The first set of equations:
For all $m \in \mathbb{N}_0$ with $0 = N - N \leqslant m$ we have
$$\tau_N^{(m)} \equiv N \equiv \tilde{\tau}_N^*.$$

Consider $n \in \mathbb{N}_0$ with $1 \leqslant n \leqslant N$ and suppose the assertion is proved for all $j \in \mathbb{N}_0$ with $n \leqslant j \leqslant N$. Consider $m \in \mathbb{N}_0$ with $N - (n-1) \leqslant m$, hence $m \geqslant 1$ and $m - 1 \geqslant N - n \geqslant 0$. Then we have

$$\begin{aligned}
C_{n-1}^{(m)} &= \left\{X_{n-1} \geqslant \tilde{Y}_{n-1}^{(m)}\right\} \cap C_{n-1}^{(0)} &&\text{by 3.6 and corresponding-assumption} \\
&= \left\{X_{n-1} \geqslant \tilde{Y}_{n-1}^{(m)}\right\} &&\text{by assumption} \\
&\subseteq \left\{X_{n-1} \geqslant \mathrm{E}\left(X_{\tau_n^{(m-1)}}\big|\mathcal{A}_{n-1}\right)\right\} &&\text{by definition} \\
&= \left\{X_{n-1} \geqslant \mathrm{E}\left(X_{\tilde{\tau}_n^*}\big|\mathcal{A}_{n-1}\right)\right\} &&\text{by ind.hyp.}
\end{aligned}$$

and by definition
$$\begin{aligned}
Y_{n-1}^* &= \max\left\{X_{n-1}, \mathrm{E}\left(Y_n^*\big|\mathcal{A}_{n-1}\right)\right\} \\
&= \max\left\{X_{n-1}, \mathrm{E}\left(\mathrm{E}\left(X_{\tilde{\tau}_n^*}\big|\mathcal{A}_n\right)\big|\mathcal{A}_{n-1}\right)\right\} \\
&= \max\left\{X_{n-1}, \mathrm{E}\left(X_{\tilde{\tau}_n^*}\big|\mathcal{A}_{n-1}\right)\right\},
\end{aligned}$$
hence
$$C_{n-1}^{(m)} \subseteq \left\{X_{n-1} = Y_{n-1}^*\right\}.$$

This shows

(3.14.2) $$C_{n-1}^{(m)} \subseteq \left\{\tilde{\tau}_{n-1}^* = n-1\right\}.$$

Due to $C_n^{(0)} = \Omega$ we have by definition
$$C_{n-1}^{(m)} = \left\{\tau_{n-1}^{(m)} = n-1\right\}$$

and hence we have, due to $\tau^{(m)}$ and τ^* being consistent,

$$\begin{aligned}
\tau_{n-1}^{(m)} &= \mathbf{1}_{C_{n-1}^{(m)}} \tau_{n-1}^{(m)} + \mathbf{1}_{\Omega \setminus C_{n-1}^{(m)}} \tau_{n-1}^{(m)} \\
&= \mathbf{1}_{C_{n-1}^{(m)}} (n-1) + \mathbf{1}_{\Omega \setminus C_{n-1}^{(m)}} \tau_{n}^{(m)} \\
&= \mathbf{1}_{C_{n-1}^{(m)}} (n-1) + \mathbf{1}_{\Omega \setminus C_{n-1}^{(m)}} \breve{\tau}_{n}^* && \text{by ind.hyp.} \\
&= \mathbf{1}_{C_{n-1}^{(m)}} \tau_{n-1}^* + \mathbf{1}_{\Omega \setminus C_{n-1}^{(m)}} \breve{\tau}_{n}^* && \text{by (3.14.2)} \\
&= \mathbf{1}_{C_{n-1}^{(m)}} \tau_{n-1}^* + \mathbf{1}_{\Omega \setminus C_{n-1}^{(m)}} \breve{\tau}_{n-1}^* \\
&= \breve{\tau}_{n-1}^*.
\end{aligned}$$

So $\breve{\tau}_{n-1}^* = \tau_{n-1}^{(m)}$ on $C_n^{(0)} = \Omega$ for all $m \in \mathbb{N}_0$ with $N - n \leq m$, which implies

$$\tau_{n-1}^{(\infty)} = \breve{\tau}_{n-1}^*. \qquad \square$$

Comment: The proof above does not work for arbitrary $C^{(0)}$, but for corresponding $C^{(0)}$ and $\tau^{(0)}$ we get similar results, mentioned in Theorem 3.18 (page 42) and proved later in Section 5 (page 51).

3.15 Definition and Remark

Now we want to consider in parallel a second instance of the algorithm with other starting parameters. To be able to distinguish it from our normal algorithm we will underline all variables of the second instance twice. We choose

$$\underline{\underline{X}} := \left(X_n \mathbf{1}_{C_n^{(0)}} \right)_{n \in \mathbb{N}_0}, \quad \underline{\underline{C^{(0)}}} := (\Omega)_{n \in \mathbb{N}_0}, \quad \underline{\underline{\tau_n^{(0)}}} := n.$$

Hence $\underline{\underline{\tau^{(0)}}}$ and $\underline{\underline{C^{(0)}}}$ are corresponding and we have by Theorem 3.14 (page 39) for all $n \in \mathbb{N}_0$ with $\underline{\underline{n \leq N}}$ and $\underline{\underline{m}} \in \mathbb{N}_0$ with $N - n \leq m$

$$\underline{\underline{\tau_n^{(m)}}} = \underline{\underline{\tau_n^{(\infty)}}} = \underline{\underline{\breve{\tau}_n^*}},$$

$$\underline{\underline{Y_n^{(m)}}} = \underline{\underline{Y_n^{(\infty)}}} = \underline{\underline{Y_n^*}}.$$

In [BKS08, Proposition 3.4] it was stated without proof that, given nonnegative X and starting with $\tau^{(0)}$ such that $\tau_n^{(0)} \in \mathcal{S}(C^{(0)})$ for each $n \in \mathbb{N}_0$ with $n \leq N$, we will end up with $\tau^{(\infty)} = \underline{\underline{\breve{\tau}^*}}$. But the following examples will show that this is wrong, as it may happen that $\underline{\underline{\breve{\tau}^*}} \notin \mathcal{S}(C^{(0)})$ despite $\tau^{(\infty)} \in \mathcal{S}(C^{(0)})$.

3.16 Example

Consider $N = 1$, $X_0 \equiv 0$, $X_1 \equiv 0$, $C_0^{(0)} = \emptyset$ and $C_1^{(0)} = \Omega$. Then we have $\underline{\underline{X}} = X$ and $\breve{\tau}_0^* \equiv 0$.

We will start with $\tau^{(0)}$ such that $\tau_n^{(0)} \in \mathcal{S}(C^{(0)})$ for each $n \in \mathbb{N}_0$ with $n \leq N$. And we end up with $\tau_0^{(\infty)} \equiv 1$, obviously $\underline{\underline{\check{\tau}_0^*}} \equiv 0 < 1 \equiv \tau_0^{(\infty)}$, hence $\tau^{(\infty)} \neq \underline{\underline{\check{\tau}^*}}$.

This example may be easily extended as follows.

3.17 Generalized Example

Consider nonnegative X and $C^{(0)}$ and $\tau^{(0)}$ such that $\tau_n^{(0)} \in \mathcal{S}(C^{(0)})$ for each $n \in \mathbb{N}_0$ with $n \leq N$. Let σ be corresponding to $C^{(0)}$.

If $\underline{\underline{\check{\tau}^*}} \lneq \sigma$, then $\tau^{(\infty)} \neq \underline{\underline{\check{\tau}^*}}$.

This is indeed true if we have

$$P\left(\bigcup_{0 \leq n < N} \bigcap_{N \geq m \geq n+1} \left\{ \underline{X_m} = 0 \right\} \setminus C_n^{(0)} \right) > 0.$$

Proof: We have $\tau^{(\infty)} = \underline{\underline{\check{\tau}^*}} \lneq \sigma \leq \tau^{(0)} \leq \tau^{(\infty)}$. □

3.18 Theorem: Algorithm Termination

- If $\tau^{(0)}$ and $C^{(0)}$ are corresponding, then

  $$\tau^{(\infty)} \text{ is (pointwise) smaller than each } \sigma \in \mathcal{T}, \text{ which is optimal in } \mathcal{S}(C^{(0)}).$$

- We always have

  $$\tau^{(\infty)} \text{ is optimal in } \mathcal{S}(C^{(0)}).$$

- If $C^{(0)}$ is essential, we even have

  $$\tau^{(\infty)} \text{ is optimal.}$$

Proof: The first statement follows due to Theorem 3.10 (page 37).
We could not prove the latter statements by backward induction. The arguments in [BKS08] use [BKS08, Proposition 3.4] which is not true as the examples above show. So mentioning these results here is in anticipation of Theorem 5.12 (page 61) and Conclusion 5.14 (page 64). □

Compared with Theorem 3.14 (page 39) the loss for essential $C^{(0)}$ (instead of $C^{(0)} \equiv \Omega$) is having not the smallest optimal stopping rule and the convergence speed has not to be that fast.

Next we examine the parameter κ. After showing by an example the influence of κ on the number of iterations, we will prove that increasing κ does never lead to worse results. Roughly spoken this is: A bigger κ leads to a smaller number of iterations, in doing so the effort of calculating the iteration increases. For the first iteration step this was mentioned in [KS06, Proposition 3.3].

3.19 Example

Consider $N = 12$ and $X \equiv (4, 3, 6, 4, 2, 3, 1, 0, 3, 5, 3, 1)$. The only optimal N-suitable family of stopping rules is $\breve{\tau}^* = (3, 3, 3, 10, 10, 10, 10, 10, 10, 10, 11, 12)$. The choice of κ has big influence on the number of iterations. Changes are shown in bold.

Given κ with

$\kappa(0) = 1$, $\kappa(1) = 1$, $\kappa(2) \leqslant 2$ we have
$$\begin{aligned}\tau^{(0)} &= (1, 2, 3, \ 4, \ 5, \ 6, \ 7, \ \ 8, \ 9, 10, 11, 12),\\ \tau^{(1)} &= (1, \mathbf{3}, 3, \ 4, \ \mathbf{6}, \ 6, \ 7, \ \mathbf{10}, \mathbf{10}, 10, 11, 12),\\ \tau^{(2)} &= (\mathbf{3}, 3, 3, \ 4, \ 6, \ 6, \mathbf{10}, \ 10, 10, 10, 11, 12),\\ \tau^{(3)} &= (3, 3, 3, \ 4, \mathbf{10}, \mathbf{10}, 10, \ 10, 10, 10, 11, 12),\\ \tau^{(4)} &= (3, 3, 3, \mathbf{10}, 10, 10, 10, \ 10, 10, 10, 11, 12) = \breve{\tau}^*,\end{aligned}$$

$\kappa(0) = 1$, $\kappa(1) = 1$, $\kappa(2) \geqslant 3$ we have
$$\begin{aligned}\tau^{(0)} &= (1, 2, 3, \ 4, \ 5, \ 6, \ 7, \ \ 8, \ 9, 10, 11, 12),\\ \tau^{(1)} &= (1, \mathbf{3}, 3, \ 4, \ \mathbf{6}, \ 6, \ 7, \ \mathbf{10}, \mathbf{10}, 10, 11, 12),\\ \tau^{(2)} &= (\mathbf{3}, 3, 3, \ 4, \ 6, \ 6, \mathbf{10}, \ 10, 10, 10, 11, 12),\\ \tau^{(3)} &= (3, 3, 3, \mathbf{10}, \mathbf{10}, \mathbf{10}, 10, \ 10, 10, 10, 11, 12) = \breve{\tau}^*,\end{aligned}$$

$\kappa(0) = 1$, $\kappa(1) = 2$ we have
$$\begin{aligned}\tau^{(0)} &= (1, 2, 3, \ 4, \ 5, \ 6, \ 7, \ \ 8, \ 9, 10, 11, 12),\\ \tau^{(1)} &= (1, \mathbf{3}, 3, \ 4, \ \mathbf{6}, \ 6, \ 7, \ \mathbf{10}, \mathbf{10}, 10, 11, 12),\\ \tau^{(2)} &= (\mathbf{3}, 3, 3, \ 4, \mathbf{10}, \mathbf{10}, \mathbf{10}, \ 10, 10, 10, 11, 12),\\ \tau^{(3)} &= (3, 3, 3, \mathbf{10}, 10, 10, 10, \ 10, 10, 10, 11, 12) = \breve{\tau}^*,\end{aligned}$$

$\kappa(0) \in \{2, 3\}$ and $\kappa(1) \leqslant 2$ we have
$$\begin{aligned}\tau^{(0)} &= (1, 2, 3, \ 4, \ 5, \ 6, \ 7, \ \ 8, \ 9, 10, 11, 12),\\ \tau^{(1)} &= (\mathbf{3}, \mathbf{3}, 3, \ 4, \ \mathbf{6}, \ 6, \mathbf{10}, \ \mathbf{10}, \mathbf{10}, 10, 11, 12),\\ \tau^{(2)} &= (3, 3, 3, \ 4, \mathbf{10}, \mathbf{10}, \mathbf{10}, \ 10, 10, 10, 11, 12),\\ \tau^{(3)} &= (3, 3, 3, \mathbf{10}, 10, 10, 10, \ 10, 10, 10, 11, 12) = \breve{\tau}^*,\end{aligned}$$

$\kappa(0) \in \{2, 3\}$ and $\kappa(1) \geqslant 3$ we have
$$\begin{aligned}\tau^{(0)} &= (1, 2, 3, \ 4, \ 5, \ 6, \ 7, \ \ 8, \ 9, 10, 11, 12),\\ \tau^{(1)} &= (\mathbf{3}, \mathbf{3}, 3, \ 4, \ \mathbf{6}, \ 6, \mathbf{10}, \ \mathbf{10}, \mathbf{10}, 10, 11, 12),\\ \tau^{(2)} &= (3, 3, 3, \mathbf{10}, \mathbf{10}, \mathbf{10}, 10, \ 10, 10, 10, 11, 12) = \breve{\tau}^*,\end{aligned}$$

$\kappa(0) \in \{4, 5\}$ we have
$$\begin{aligned}\tau^{(0)} &= (1, 2, 3, \ 4, \ 5, \ 6, \ 7, \ \ 8, \ 9, 10, 11, 12),\\ \tau^{(1)} &= (\mathbf{3}, \mathbf{3}, 3, \ 4, \mathbf{10}, \mathbf{10}, \mathbf{10}, \ \mathbf{10}, \mathbf{10}, 10, 11, 12),\\ \tau^{(2)} &= (3, 3, 3, \mathbf{10}, 10, 10, 10, \ 10, 10, 10, 11, 12) = \breve{\tau}^*,\end{aligned}$$

$\kappa(0) \geqslant 6$ we have
$$\begin{aligned}\tau^{(0)} &= (1, 2, 3, \ 4, \ 5, \ 6, \ 7, \ \ 8, \ 9, 10, 11, 12),\\ \tau^{(1)} &= (\mathbf{3}, \mathbf{3}, 3, \mathbf{10}, \mathbf{10}, \mathbf{10}, \mathbf{10}, \ \mathbf{10}, \mathbf{10}, 10, 11, 12) = \breve{\tau}^*.\end{aligned}$$

3.20 Lemma

Consider $\sigma, \tau \in \mathcal{T}$ with $\tau \leqslant \sigma$ and $i, p \in \mathbb{N}_0$ with $i \leqslant p$. Then we have
$$X_{\sigma_i} = X_{\sigma_p} \text{ on } \{\tau_i = p\}.$$

Proof: By the definition of $\mathcal{T}$ in Definition 1.10 (page 12) we have
$$\{\sigma_n > n\} \subseteq \{\sigma_n = \sigma_{n+1}\} \text{ for all } n \in \mathbb{N}_0 \text{ with } n < N,$$
hence it follows by induction
$$\{\sigma_i \geqslant p\} \subseteq \{\sigma_i = \sigma_p\}.$$
Due to $\tau \leqslant \sigma$ we have $\sigma_i \geqslant p$ on $\{\tau_i = p\}$ and hence $\sigma_i = \sigma_p$ on $\{\tau_i = p\}$. □

3.21 Theorem: Influence of κ

Assume $C^{(0)} \equiv \Omega$. Consider a second instance of the algorithm, where all variables are marked with a $'$, with identical input variables except for κ' which is only assumed to be nowhere smaller than κ. We have

$$\hat{Y}_n^{(m)} \leqslant \hat{Y}'^{(m)}_n,$$
$$\check{Y}_n^{(m)} \leqslant \check{Y}'^{(m)}_n,$$
$$C_n^{(m)} \supseteq C'^{(m)}_n,$$
$$\hat{C}_n^{(m)} \supseteq \hat{C}'^{(m)}_n,$$
$$\Theta_n^{(m)} \supseteq \Theta'^{(m)}_n,$$
$$\hat{\Theta}_n^{(m)} \supseteq \hat{\Theta}'^{(m)}_n,$$
$$\tau_n^{(m)} \leqslant \tau'^{(m)}_n,$$
$$\hat{\tau}_n^{(m)} \leqslant \hat{\tau}'^{(m)}_n,$$
$$Y_i^{(m)} \leqslant Y'^{(m)}_i.$$

Proof: We will show this by induction. The initial assumption of this theorem will not be necessary for $m = 0$ but in the induction step in (3.21.2).

Since we use the same input variables the induction hypothesis is true for $m = 0$. Consider $m \in \mathbb{N}$ and the induction hypothesis being true for $m - 1$. Then it is true for m by definition in Part 2.13 (page 22) except for the last inequality. Since $\tau_n^{(m)} \leqslant \tau'^{(m)}_n$ for all $n \in \mathbb{N}_0$ with $n \leqslant N$ we have by Lemma 3.20 (page 44)

(3.21.1) $\qquad X_{\tau_i^{'(m)}} = X_{\tau_p^{'(m)}}$ on $\left\{\tau_i^{(m)} = p\right\}$ for all $i, p \in \mathbb{N}_0$ with $i \leqslant p \leqslant N$

By Lemma 3.2 (page 30) and Conclusion 3.4 (page 31) we have for all $p \in \mathbb{N}_0$ with $p \leqslant N$

(3.21.2) $\qquad X_p \leqslant Y'^{(1)}_p \leqslant Y'^{(m)}_p = \mathrm{E}\left(X_{\tau_p^{'(m)}} \big| \mathcal{A}_p\right)$ on $C_p^{(0)} = \Omega.$

Thus we have for all $i \in \mathbb{N}_0$ with $i \leq N$

$$
\begin{aligned}
Y_i'^{(m)} &\stackrel{\text{def.}}{=} \mathrm{E}\left(X_{\tau_i'^{(m)}}\Big|\mathcal{A}_i\right) \\
&= \mathrm{E}\left(\sum_{p=i}^{N}\mathbf{1}_{\{\tau_i^{(m)}=p\}}X_{\tau_i'^{(m)}}\Big|\mathcal{A}_i\right) \\
&\stackrel{(3.21.1)}{=} \mathrm{E}\left(\sum_{p=i}^{N}\mathbf{1}_{\{\tau_i^{(m)}=p\}}X_{\tau_p'^{(m)}}\Big|\mathcal{A}_i\right) \\
&= \mathrm{E}\left(\sum_{p=i}^{N}\mathbf{1}_{\{\tau_i^{(m)}=p\}}\mathrm{E}\left(X_{\tau_p'^{(m)}}\Big|\mathcal{A}_p\right)\Big|\mathcal{A}_i\right) \\
&\stackrel{(3.21.2)}{\geq} \mathrm{E}\left(\sum_{p=i}^{N}\mathbf{1}_{\{\tau_i^{(m)}=p\}}X_p\Big|\mathcal{A}_i\right) \\
&= \mathrm{E}\left(X_{\tau_i^{(m)}}\Big|\mathcal{A}_i\right) \\
&\stackrel{\text{def.}}{=} Y_i^{(m)}. \qquad \square
\end{aligned}
$$

4 Different Finite Time Horizons Compared

4.1 Idea

As written in Convention 1.9 (page 11) nearly all variables depend on N. For easier reading we mostly did not include this dependence in our notations. Now it will be important to have a look at the number of steps N, so whenever necessary we shall write N as an additional inner subscript. For example we will write $C_{N,k}^{(0)}$ instead of $C_k^{(0)}$.

4.2 Example

A first idea was that we might have $\tau_{N,n}^{(m)} = \tau_{\infty,n}^{(m)} \wedge N$ for m, N, n, at least under reasonable assumptions. But this does not hold in general as the following example shows:

Example: Consider $N_1, N_2 \in \mathbb{N} \cup \{\infty\}$ with $N_1 < N_2$.
Consider $C_{N_i,n}^{(0)} = \Omega$ for all $n \in \mathbb{N}_0$ and $i \in \{1,2\}$.
Consider a process X given for all $n \in \mathbb{N}$ by

$$X_n = \begin{cases} 1 & \text{if } n < N_1, \\ 0 & \text{if } n = N_1, \\ 2 & \text{if } n > N_1. \end{cases}$$

Then we have for all $i \in \{1,2\}$ and $n \in \mathbb{N}_0$ with $n \leqslant N_i$

$$\tau_{N_i,n}^{(0)} = n$$

and

$$Y_{N_i,n}^{(0)} = X_n = \begin{cases} 1 & \text{if } n < N_1, \\ 0 & \text{if } n = N_1, \\ 2 & \text{if } n > N_1. \end{cases}$$

For all $m \in \mathbb{N}$ and $n \in \mathbb{N}_0$ with $n \leqslant N_1$ we have

$$\tilde{Y}_{N_1,n}^{(m)} = \begin{cases} 1 & \text{if } n < N_1, \\ 0 & \text{otherwise.} \end{cases}$$

For all $m \in \mathbb{N}$ and $n \in \mathbb{N}_0$ with $n \leqslant N_2$

$$\tilde{Y}_{N_2,n}^{(m)} = \begin{cases} 1 & \text{if } n \leqslant N_1 - \sum_{k=0}^{m-1} \kappa(k), \\ 2 & \text{otherwise.} \end{cases}$$

So we have for all $m \in \mathbb{N}$ and $n \in \mathbb{N}_0$ with $n \leq N_1$
$$C^{(m)}_{N_1,n} = \Omega,$$
and for all $m \in \mathbb{N}$ and $n \in \mathbb{N}_0$ with $n \leq N_2$
$$C^{(m)}_{N_2,n} = \begin{cases} \emptyset & \text{if } N_1 - \sum_{k=0}^{m-1} \kappa(k) < n \leq N_1, \\ \Omega & \text{otherwise.} \end{cases}$$

Thus we have for all $m \in \mathbb{N}$ and $n \in \mathbb{N}_0$ with $n \leq N_1$
$$\tau^{(m)}_{N_1,n} = n,$$
and for all $m \in \mathbb{N}$ and $n \in \mathbb{N}_0$ with $n \leq N_2$
$$\tau^{(m)}_{N_2,n} = \begin{cases} N_1 + 1 & \text{if } N_1 - \sum_{k=0}^{m-1} \kappa(k) < n \leq N_1, \\ n & \text{otherwise.} \end{cases}$$

And we have for all $m \in \mathbb{N}$ and $n \in \mathbb{N}_0$ with $n \leq N_1$
$$Y^{(0)}_{N_1,n} = \begin{cases} 1 & \text{if } n < N_1, \\ 0 & \text{if } n = N_1, \end{cases}$$
and for all $m \in \mathbb{N}$ and $n \in \mathbb{N}_0$ with $n \leq N_2$
$$Y^{(0)}_{N_2,n} = \begin{cases} 2 & \text{if } n > N_1 - \sum_{k=0}^{m-1} \kappa(k), \\ 1 & \text{otherwise.} \end{cases}$$

So it follows that for all $m \in \mathbb{N}$ and for all $n \in \mathbb{N}_0$ with $N_1 - \sum_{k=0}^{m-1} \kappa(k) \leq n < N_1$
$$\tau^{(m)}_{N_1,n} = n < N_1 = \tau^{(m)}_{N_2,n} \wedge N_1. \qquad \square$$

4.3 Lemma

Let $N_1, N_2 \in \mathbb{N} \cup \{\infty\}$ with $N_1 < N_2$, $m \in \mathbb{N}_0$ with

(4.3.1) $\qquad m = 0 \implies X_n \leq Y^{(0)}_{N_2,n}$ on $\left\{\tau^{(0)}_{N_1,n} = n\right\}$ for all $n \in \mathbb{N}_0$ with $n \leq N_1$

and

(4.3.2) $\qquad m = 1 \implies \tau^{(0)}_{N_2,n} \equiv n$ for all $n \in \mathbb{N}_0$ with $n \leq N_1$.

Assume $C^{(0)}_{N_2} \equiv \Omega$.
Suppose that
$$\tau^{(m)}_{N_1,n} \leq \tau^{(m)}_{N_2,n} \text{ for all } n \in \mathbb{N}_0 \text{ with } n \leq N_1.$$
Then we have
$$Y^{(m)}_{N_1,n} \leq Y^{(m)}_{N_2,n} \text{ for all } n \in \mathbb{N}_0 \text{ with } n \leq N_1.$$
Remark: The proof given below does not work for arbitrary $C^{(0)}_{N_2}$.

Proof by Backward Induction:
We have
$$Y^{(m)}_{N_1,N_1} = \mathrm{E}\left(X_{\tau^{(m)}_{N_1,N_1}}\Big|\mathcal{A}_{N_1}\right) = X_{N_1} \underset{3.7}{\overset{\text{ass.}}{\leqslant}} Y^{(m)}_{N_2,N_1} \text{ on } C^{(0)}_{N_2,N_1} = \Omega.$$

Consider some $n \in \mathbb{N}_0$ with $n < N_1$ and $Y^{(m)}_{N_1,n+1} \leqslant Y^{(m)}_{N_2,n+1}$.
On the set $\left\{\tau^{(m)}_{N_1,n} = n\right\}$ we have

$$Y^{(m)}_{N_1,n} = X_n \overset{(4.3.2),3.7}{\underset{(4.3.1)}{\leqslant}} Y^{(m)}_{N_2,n} \text{ on } C^{(0)}_{N_2,n} = \Omega.$$

For all $i \in \{1,2\}$ we have

(4.3.3) $$\left\{\tau^{(m)}_{N_i,n} > n\right\} \subseteq \left\{\tau^{(m)}_{N_i,n} = \tau^{(m)}_{N_i,n+1}\right\}$$

and thus on $\left\{\tau^{(m)}_{N_i,n} > n\right\}$

$$\begin{aligned}
Y^{(m)}_{N_i,n} &= \mathrm{E}\left(X_{\tau^{(m)}_{N_i,n}}\Big|\mathcal{A}_n\right) && \text{by def.}\\
&= \mathrm{E}\left(X_{\tau^{(m)}_{N_i,n+1}}\Big|\mathcal{A}_n\right) && \text{by (4.3.3)}\\
&= \mathrm{E}\left(\mathrm{E}\left(X_{\tau^{(m)}_{N_i,n+1}}\Big|\mathcal{A}_{n+1}\right)\Big|\mathcal{A}_n\right)\\
&= \mathrm{E}\left(Y^{(m)}_{N_i,n+1}\Big|\mathcal{A}_n\right) && \text{by def.}
\end{aligned}$$

So we have on $\left\{\tau^{(m)}_{N_1,n} > n\right\} \subseteq \left\{\tau^{(m)}_{N_2,n} > n\right\}$

$$Y^{(m)}_{N_1,n} = \mathrm{E}\left(Y^{(m)}_{N_1,n+1}\Big|\mathcal{A}_n\right) \overset{\text{ind.ass.}}{\leqslant} \mathrm{E}\left(Y^{(m)}_{N_2,n+1}\Big|\mathcal{A}_n\right) = Y^{(m)}_{N_2,n}. \qquad \square$$

4.4 Lemma

Consider $N_1, N_2 \in \mathbb{N} \cup \{\infty\}$ with $N_1 < N_2$.
Assume $C^{(0)}_{N_2} = C^{(0)}_{N_1} \equiv \Omega$.
Consider $m \in \mathbb{N}_0$ satisfying assumptions (4.3.1) (page 48) and (4.3.2) (page 48).
Suppose that for all $n \in \mathbb{N}_0$ with $n \leqslant N_1$

$$\tau^{(m)}_{N_1,n} \leqslant \tau^{(m)}_{N_2,n}.$$

Then we have for all $n \in \mathbb{N}_0$ with $n \leqslant N_1$

$$\begin{aligned}
Y^{(m)}_{N_1,n} &\leqslant Y^{(m)}_{N_2,n},\\
\tilde{Y}^{(m)}_{N_1,n} &\leqslant \tilde{Y}^{(m)}_{N_2,n},\\
C^{(m+1)}_{N_2,n} &\subseteq C^{(m+1)}_{N_1,n},\\
\tau^{(m+1)}_{N_1,n} &\leqslant \tau^{(m+1)}_{N_2,n}.
\end{aligned}$$

Proof: The first inequality follows by Lemma 4.3 (page 48).
For the other inequalities define for all $i \in \{1,2\}$ and for all $n \in \mathbb{N}_0$ with $n \leq N_1$

$$A_{N_i,n}^{(m)} := \{p \in \mathbb{N}_0 \; ; \; n \leq p < \min\{n + \kappa(m), N_i\} + 1\}.$$

Thus we have for all $n \in \mathbb{N}_0$ with $n \leq N_1$ and for all $p \in A_{N_1,n}^{(m)}$

$$\mathrm{E}\left(Y_{N_1,p}^{(m)} \middle| \mathcal{A}_n\right) \leq \mathrm{E}\left(Y_{N_2,p}^{(m)} \middle| \mathcal{A}_n\right).$$

Furthermore we have for all $n \in \mathbb{N}_0$ with $n \leq N_1$

$$A_{N_1,n}^{(m)} \subseteq A_{N_2,n}^{(m)}.$$

So we have for all $n \in \mathbb{N}_0$ with $n \leq N_1$

$$\widetilde{Y}_{N_1,n}^{(m+1)} = \sup_{p \in A_{N_1,n}^{(m)}} \mathrm{E}\left(Y_{N_1,p}^{(m)} \middle| \mathcal{A}_n\right) \leq \sup_{p \in A_{N_1,n}^{(m)}} \mathrm{E}\left(Y_{N_2,p}^{(m)} \middle| \mathcal{A}_n\right)$$

$$\leq \sup_{p \in A_{N_2,n}^{(m)}} \mathrm{E}\left(Y_{N_2,p}^{(m)} \middle| \mathcal{A}_n\right) = \widetilde{Y}_{N_2,n}^{(m+1)},$$

thus we have

$$\left\{\widetilde{Y}_{N_1,n}^{(m+1)} \leq X_n\right\} \subseteq \left\{\widetilde{Y}_{N_2,n}^{(m+1)} \leq X_n\right\},$$

hence

$$C_{N_2,n}^{(m+1)} \subseteq C_{N_1,n}^{(m+1)}.$$

This implies for all $n \in \mathbb{N}_0$ with $n \leq N_1$

$$\tau_{N_1,n}^{(m+1)} \leq \tau_{N_2,n}^{(m+1)}. \qquad \square$$

4.5 Theorem

Consider $N_1, N_2 \in \mathbb{N} \cup \{\infty\}$ with $N_1 < N_2$.
Assume $C_{N_2}^{(0)} = C_{N_1}^{(0)} \equiv \Omega$, $\tau_{N_1}^{(0)}$ and $\tau_{N_2}^{(0)}$ determined by $C^{(0)}$ and

$$X_n \leq Y_{N_2,n}^{(0)} \text{ on } \left\{\tau_{N_1,n}^{(0)} = n\right\} \text{ for all } n \in \mathbb{N}_0 \text{ with } n \leq N_1.$$

Then we have for all $m \in \mathbb{N}_0$ and for all $n \in \mathbb{N}_0$ with $n \leq N_1$

$$Y_{N_1,n}^{(m)} \leq Y_{N_2,n}^{(m)} \text{ on } C_{N_2,n}^{(0)},$$
$$\widetilde{Y}_{N_1,n}^{(m)} \leq \widehat{Y}_{N_2,n}^{(m)} \text{ on } C_{N_2,n}^{(0)},$$
$$C_{N_2,n}^{(m)} \subseteq C_{N_1,n}^{(m)},$$
$$\tau_{N_1,n}^{(m)} \leq \tau_{N_2,n}^{(m)}.$$

Proof: Proved by induction, using Lemma 4.4 (page 49). $\qquad \square$

5 General Case

The FII Algorithm introduced in Section 2 (page 15) gives optimal results for the finite time horizon as seen in Section 3 (page 27). In this section we show this for the infinite time horizon as well. The techniques within the proofs will be different since backward induction is not available here.

First we will repeat the result of [Irl80] in our nomenclature.

5.1 Theorem from [Irl80] for $\kappa \equiv 1$

Assume $C^{(0)} \equiv \Omega$, $\kappa \equiv 1$ and
$$E(X_\tau) \text{ exists for all } \tau \in \mathcal{S},$$
$$E\left(X_{\sup\{\sigma_n\,;\,n \in \mathbb{N}_0\}} \middle| \mathcal{A}_0\right) = \lim_{n \to \infty} E(X_{\sigma_n} | \mathcal{A}_0) \text{ for all } \sigma \in \mathcal{S}^{\mathbb{N}_0} \text{ non-decreasing.}$$
Then $\mathcal{S}_0^\infty = \mathcal{S}$ and
$$\tau_0^{(\infty)} \text{ is optimal in } \mathcal{S}$$
$$\tau_0^{(m)} \leqslant \tau_0^{(m+1)} \text{ for all } m \in \mathbb{N}_0,$$
$$Y_0^{(m)} \leqslant Y_0^{(m+1)} \text{ for all } m \in \mathbb{N}_0.$$

Thus it follows
$$Y_0^{(\infty)} \stackrel{\text{def.}}{\underset{\text{incr.}}{=}} \lim_{m \to \infty} Y_0^{(m)}$$
$$\stackrel{\text{def.}}{=} \lim_{m \to \infty} E\left(X_{\tau_0^{(m)}} \middle| \mathcal{A}_0\right) \stackrel{\text{ass.}}{=} E\left(X_{\lim_{m \to \infty} \tau_0^{(m)}} \middle| \mathcal{A}_0\right) \stackrel{\text{def.}}{\underset{\text{incr.}}{=}} E\left(X_{\tau_0^{(\infty)}} \middle| \mathcal{A}_0\right).$$

If
$$C^{(l)} = C^{(l+1)} \text{ for some } l \in \mathbb{N}_0,$$
then obviously
$$C^{(n+l)} = C^{(l)} = C^{(\infty)} \text{ for all } n \in \mathbb{N}_0,$$
$$\tau_0^{(n+l)} = \tau_0^{(l)} = \tau_0^{(\infty)} \text{ for all } n \in \mathbb{N}_0.$$

Proof: A direct proof can be found in [Irl80, Part 2, pages 180-182], a proof for a more general situation follows within this section. □

We will now expand this theorem to arbitrary κ (instead of $\kappa \equiv 1$) and arbitrary timepoint j (instead of looking only at time point 0), see Main Theorem 5.3 (page 52). In Chapter 9 (page 89) we will see that not using $\kappa \equiv 1$ may increase the speed of the algorithm. The general assumption made above in [Irl80] as well as in [Irl06] and [Irl09] has to be slightly enlarged for our purposes to the General Assumption 5.2 (page 52).

5.2 General Assumption

Assume

(5.2.1) $$E(X_\tau) \text{ exists for all } \tau \in \mathcal{S},$$

(5.2.2) $$\lim_{n\to\infty} E(X_{\sigma_n}|\mathcal{A}_j) = E\left(X_{\lim_{n\to\infty}\sigma_n}\Big|\mathcal{A}_j\right)$$
for all $j \in \mathbb{N}_0, \sigma \in \mathcal{S}^{\mathbb{N}_0}$ non-decreasing.

5.3 Main Theorem

$$\tau^{(\infty)} \text{ is optimal in } \mathcal{S}(C^{(0)})$$

and for all $j \in \mathbb{N}_0, m \in \mathbb{N}$ we have
$$\tau_j^{(m)} \leq \tau_j^{(m+1)},$$
$$Y_j^{(m)} \leq Y_j^{(m+1)}.$$

If we have $\tau^{(0)} \in \mathcal{S}(C^{(0)})$, this is also true for $m = 0$.

We have for all $j \in \mathbb{N}_0$
$$Y_j^{(\infty)} = E\left(X_{\tau_j^{(\infty)}}\Big|\mathcal{A}_j\right).$$

If
$$C^{(l)} = C^{(l+1)} \qquad \text{for some } l \in \mathbb{N}_0,$$

then we have

$$C^{(l)} = C^{(l+n)} = C^{(\infty)} \text{ for all } n \in \mathbb{N}_0,$$
$$\tau^{(l)} = \tau^{(l+n)} = \tau^{(\infty)} \text{ for all } n \in \mathbb{N}_0,$$
$$\Theta^{(l)} = \Theta^{(l+n)} = \Theta^{(\infty)} \text{ for all } n \in \mathbb{N}_0,$$
$$Y^{(l)} = Y^{(l+n)} = Y^{(\infty)} \text{ for all } n \in \mathbb{N}_0.$$

If $C^{(0)} \equiv \Omega$ or $C^{(0)}$ is essential, then

$$\tau^{(\infty)} \text{ is optimal.}$$

5.4 Delimitating Example

If we have $\tau^{(0)} \notin \mathcal{S}(C^{(0)})$, then $\tau_j^{(0)} \leq \tau_j^{(1)}$ may be wrong for some $j \in \mathbb{N}_0$.

Proof: See [BKS08, Example 3.3]. □

We now give a reason why it has not been necessary to mention the General Assumption 5.2 (page 52) before with finite N in Remark 5.5 (page 53) and Remark 5.6 (page 53). We will further discuss the General Assumption 5.2 (page 52) in Lemma 5.7 (page 53) and Remark 5.8 (page 54).

The nomenclature used in this section is mainly defined in Definition 2.1 (page 15) and is then enlarged in Definition 5.9 (page 54). We then give the proof of Main Theorem 5.3 (page 52) in several steps. These are the necessary Lemma 5.11 (page 55), followed by the very detailed Theorem 5.12 (page 61), whereafter the latter parts are proved in Conclusion 5.13 (page 63), Conclusion 5.14 (page 64) and Remark 5.15 (page 64).

5.5 Remark: Limites

Consider a non-decreasing $\sigma \in \mathcal{S}^{\mathbb{N}_0}$. We have
$$\lim_{n \to \infty} X_{\sigma_n} = X_{\lim_{n \to \infty} \sigma_n} \text{ on } \{\sup\{\sigma_n \; ; \; n \in \mathbb{N}_0\} < \infty\}.$$
If $X_\infty = \lim_{n \to \infty} X_n$ then we have
$$\lim_{n \to \infty} X_{\sigma_n} = X_{\lim_{n \to \infty} \sigma_n}.$$

5.6 Remarks

- If N is finite, we set $X_{N+n} = X_N$ for all $n \in \mathbb{N}$ and $X_\infty = X_N$ (see Remark 2.20 (page 26)). Hence we have
$$X_\infty = \lim_{n \to \infty} X_n$$
and for all non-decreasing $\sigma \in \mathcal{S}^{\mathbb{N}_0}$ there is non-decreasing $\tau \in \left(\mathcal{S}_0^N\right)^{\mathbb{N}_0}$ with $X_{\sigma_n} = X_{\tau_n}$ for all $n \in \mathbb{N}$. Thus we have by Lemma 3.12 (page 38)
$$\lim_{n \to \infty} \mathrm{E}\left(X_{\sigma_n} | \mathcal{A}_j\right) = \mathrm{E}\left(X_{\lim_{n \to \infty} \sigma_n} \Big| \mathcal{A}_j\right)$$
for all $j \in \mathbb{N}_0, \sigma \in \mathcal{S}^{\mathbb{N}_0}$ non-decreasing.

- This is not generally true for infinite N. Consider the following example: For all $n \in \mathbb{N}_0$ let $X_n = n \mathbf{1}_{]0, \frac{1}{n}]}$. Then we have
$$X_\infty = \lim_{n \to \infty} X_n = 0.$$
Now consider $\sigma \in \mathcal{S}^{\mathbb{N}_0}$ with $\sigma_n = n$ for all $n \in \mathbb{N}_0$. Then we have for all $j \in \mathbb{N}_0$
$$\mathrm{E}\left(X_{\lim_{n \to \infty} \sigma_n} \Big| \mathcal{A}_j\right) = \mathrm{E}(X_\infty | \mathcal{A}_j) = 0 \neq 1 = \lim_{n \to \infty} \mathrm{E}(X_n | \mathcal{A}_j) = \lim_{n \to \infty} \mathrm{E}(X_{\sigma_n} | \mathcal{A}_j).$$

5.7 Lemma

A sufficient condition for (5.2.1) and (5.2.2) is

(5.7.1) $$\mathrm{E}\left(\sup_{n \in \mathbb{N}} |X_n|\right) < \infty \text{ and } X_\infty = \lim_{n \to \infty} X_n.$$

Proof: Assume (5.7.1). We have for each $\tau \in \mathcal{S}$
$$|X_\tau| \leqslant \sup_{n \in \mathbb{N}} |X_n|,$$
where the rhs is an integrable random variable, hence we have (5.2.1).

Now consider $j \in \mathbb{N}_0, \sigma \in \mathcal{S}^{\mathbb{N}_0}$ non-decreasing.

By Conditional Lebesgue Dominated Convergence we have
$$\lim_{n \to \infty} \mathrm{E}\left(X_{\sigma_n}\middle|\mathcal{A}_j\right) = \mathrm{E}\left(\lim_{n \to \infty} X_{\sigma_n}\middle|\mathcal{A}_j\right)$$
and by Remark 5.5 (page 53) we have
$$\mathrm{E}\left(\lim_{n \to \infty} X_{\sigma_n}\middle|\mathcal{A}_j\right) = \mathrm{E}\left(X_{\lim_{n \to \infty} \sigma_n}\middle|\mathcal{A}_j\right)$$
and thus (5.2.2) holds. □

5.8 Remarks

- If N is finite and embedded in the infinite case as written in Remark 2.20 (page 26), then we have (5.2.2) by Remark 5.6 (page 53) and (5.2.1) by Lemma 1.4 (page 10).

- The assumption (5.2.2) is equivalent to the following assumption:
$$\lim_{n \to \infty} \int_G X_{\sigma_n} dP = \int_G X_{\lim_{n \to \infty} \sigma_n} dP$$
for all $j \in \mathbb{N}_0$, $G \in \mathcal{A}_j$, $\sigma \in \mathcal{S}^{\mathbb{N}_0}$ non-decreasing.

- Due to the assumption (5.2.1) we have $\mathcal{S} = \mathcal{S}_0^*$ (see Definition 1.3 (page 9)).

5.9 Definitions

For all $C \in \mathcal{C}$ and $\sigma \in \mathcal{T}$ define
$$\tau_\sigma(C) := \inf\{k \geqslant \sigma \,;\, \mathbf{1}_{C_k} = 1\}.$$
For all $n \in \mathbb{N}_0$ write $\tau_n(C)$ for $\tau_\sigma(C)$ with $\sigma \equiv n$.

For all $C \in \mathcal{C}$ and $\kappa \in \mathbb{N} \cup \{\infty\}$ define $C^{*\kappa}$ by

(5.9.1) $\qquad C_n^{*\kappa} := C_n \cap \bigcap_{j=1}^{\kappa} \{X_n \geqslant \mathrm{E}(X_{\tau_{n+j}(C)}|\mathcal{A}_n)\}$ for all $n \in \mathbb{N}_0$, $C_\infty^{*\kappa} := \Omega$.

For all $C \in \mathcal{C}$, $\kappa \in \mathbb{N} \cup \{\infty\}$, $\rho \in \mathcal{S}(C)$ with $\sigma \leqslant \rho \leqslant \tau_\sigma(C^{*\kappa})$ define
$$\widehat{\rho}^\kappa := \rho \mathbf{1}_{\{\rho = \tau_\sigma(C^{*\kappa})\}} + \sum_{n=0}^{\infty} \sum_{j=1}^{\kappa} \mathbf{1}_{\{\rho=n\} \cap \{j=\min\{l \in \mathbb{N};\, X_n < \mathrm{E}(X_{\tau_{n+l}(C)}|\mathcal{A}_n)\}\}} \left(\tau_{n+j}(C) \wedge \tau_\sigma(C^{*\kappa})\right).$$
If ambiguity is impossible, we do not write the index κ.

The improvement step in (5.9.1) coincides with that in Algorithm 2.13 (page 22) due to (2.16.4).

5.10 Lemma: Optimality

Consider $C \in \mathcal{C}$ with $\tau(C)$ being optimal in $\mathcal{S}(C)$. Then for each $\kappa \in \mathbb{N} \cup \{\infty\}$ we have $C = C^{*\kappa}$.

Proof: We have for all $j \in \mathbb{N}$ $\mathcal{T}_{n+j}(C) \in \mathcal{S}(C) \cap \mathcal{S}_{n+j}^N \subseteq \mathcal{S}(C) \cap \mathcal{S}_n^N$ and hence by Definition 1.11 (page 12)
$$X_n = \mathrm{E}\left(X_{\tau_n(C)}\big|\mathcal{A}_n\right) = \mathrm{E}\left(X_{\tau_{n+j}(C)}\big|\mathcal{A}_n\right) \text{ on } C_n.$$
The assertion follows by the definition of $C^{*\kappa}$. □

We make adaptions of the ideas in [Irl80, Part 2, Theorem 2.1, Proof part (i), page 181] to conditional expectations and arbitrary κ. The idea for arbitrary κ, but with finite N, can be found in [KS06] and [BS06] and has been extended in Section 3 (page 27) where backward induction was the main tool of the proofs.

5.11 Lemma: One Step Improvement

Suppose $C \in \mathcal{C}$, $i \in \mathbb{N}_0$, $\kappa \in \mathbb{N} \cup \{\infty\}$, and $\sigma \in \mathcal{S}(C)$ with

(5.11.1) $$\sigma \geqslant i.$$

For all

(5.11.2) $$\rho \in \mathcal{S}(C) \text{ with } \sigma \leqslant \rho \leqslant \tau_\sigma(C^*)$$

we have

(5.11.3) $$\hat{\rho} \in \mathcal{S}(C) \text{ with } \sigma \leqslant \hat{\rho} \leqslant \tau_\sigma(C^*),$$
(5.11.4) $$\rho \leqslant \hat{\rho},$$
(5.11.5) $$\rho + 1 \leqslant \hat{\rho} \text{ on } \{\rho < \tau_\sigma(C^*)\},$$
(5.11.6) $$\mathrm{E}\left(X_\rho\big|\mathcal{A}_i\right) \leqslant \mathrm{E}\left(X_{\hat{\rho}}\big|\mathcal{A}_i\right),$$

and using assumption (5.2.2) (page 52) also

(5.11.7) $$\mathrm{E}\left(X_\rho\big|\mathcal{A}_i\right) \leqslant \mathrm{E}\left(X_{\tau_\sigma(C^*)}\big|\mathcal{A}_i\right).$$

Remark: If $\kappa = 1$, then

(5.11.8) $$C_n^* = C_n \cap \left\{X_n \geqslant \mathrm{E}\left(X_{\tau_{n+1}(C)}\big|\mathcal{A}_n\right)\right\} \text{ for all } n \in \mathbb{N}_0,$$
$$\hat{\rho} = \rho \mathbf{1}_{\{\rho = \tau_\sigma(C^*)\}} + \sum_{n=i}^{\infty} \mathbf{1}_{\{\rho=n\} \cap \left\{\mathrm{E}\left(X_{\tau_{n+1}(C)}\big|\mathcal{A}_n\right) > X_n\right\}} \tau_{n+1}(C).$$

We will have much shorter proofs for $\kappa = 1$ then for arbitrary κ. We included both proofs for comparance in the sequel.

Proof of (5.11.7) using (5.11.3) to (5.11.6):
Define $\sigma_0 := \rho$ and inductively

$$\sigma_{k+1} := \hat{\sigma}_k \text{ for all } k \in \mathbb{N}_0.$$

Now we will prove by induction that

(5.11.9) $$\rho \leqslant \sigma_k \leqslant \sigma_{k+1} \leqslant \tau_\sigma(C^*) \text{ for all } k \in \mathbb{N}_0$$

We have $\sigma_0 = \rho \leq \tau_\sigma(C^*)$, hence we have by (5.11.3) and (5.11.4) $\sigma_0 \leq \sigma_1 \leq \tau_\sigma(C^*)$. Consider $k \in \mathbb{N}_0$ with $\sigma_k \leq \tau_\sigma(C^*)$. It follows $\sigma_k \leq \sigma_{k+1} \leq \tau_\sigma(C^*)$ by the same equations.

We have (by induction) for all $k \in \mathbb{N}$

(5.11.10) $$\mathrm{E}\left(X_{\sigma_0}|\mathcal{A}_i\right) \stackrel{(5.11.6)}{\leq} \mathrm{E}\left(X_{\sigma_k}|\mathcal{A}_i\right).$$

By (5.11.9) and (5.11.5) we have
$$\lim_{k \to \infty} \sigma_k \leq \tau_\sigma(C^*)$$
and
$$\sigma_k + 1 \leq \sigma_{k+1} \text{ on } \{\sigma_k < \tau_\sigma(C^*)\} \text{ for all } k \in \mathbb{N}.$$

So we have

(5.11.11) $$\sigma_0 \leq \sigma_1 \leq \sigma_2 \leq \ldots \leq \lim_{k \to \infty} \sigma_k = \tau_\sigma(C^*).$$

It follows
$$\begin{aligned}
\mathrm{E}\left(X_\rho|\mathcal{A}_i\right) &= \mathrm{E}\left(X_{\sigma_0}|\mathcal{A}_i\right) && \text{by def.} \\
&\leq \lim_{k \to \infty} \mathrm{E}\left(X_{\sigma_k}|\mathcal{A}_i\right) && \text{by (5.11.10)} \\
&= \mathrm{E}\left(X_{\lim_{k \to \infty} \sigma_k}|\mathcal{A}_i\right) && \text{by (5.2.2) (page 52)} \\
&= \mathrm{E}\left(X_{\tau_\sigma(C^*)}|\mathcal{A}_i\right) && \text{by (5.11.11).}
\end{aligned}$$
$\square$

Proof of (5.11.3) to (5.11.6) for $\kappa = 1$:
(This is a special version for $\kappa = 1$ of the arbitrary case below and due to this much shorter.)
Obviously we have

(5.11.12) $$\rho = \hat{\rho} = \tau_\sigma(C^*) \text{ on } \{\rho = \tau_\sigma(C^*)\} \stackrel[(5.11.1)]{(5.11.2)}{=} \{\rho = \infty\} \cup \bigcup_{n=i}^{\infty} \{\rho = n\} \cap C_n^*.$$

For all $n \in \mathbb{N}_0$ with $i \leq n$ we have

$$\rho < \rho + 1 = n + 1 \leq \overbrace{\tau_{n+1}(C)}^{=\hat{\rho}} \leq \tau_\sigma(C^*)$$
$$\text{on } \{\rho = n\} \cap \{\mathrm{E}\left(X_{\tau_{n+1}(C)}|\mathcal{A}_n\right) > X_n\}.$$

Furthermore we have (where $\uplus$ is here and in the following used for disjoint unions)

(5.11.13) $$\{\rho < \tau_\sigma(C^*)\} \stackrel{(5.11.12)}{=} \biguplus_{n=i}^{\infty} \{\rho = n\} \cap \{\mathrm{E}\left(X_{\tau_{n+1}(C)}|\mathcal{A}_n\right) > X_n\}.$$

For (5.11.6) it suffices to prove for all $G \in \mathcal{A}_i$
$$\int_G X_\rho \, \mathrm{d}P \leq \int_G X_{\hat{\rho}} \, \mathrm{d}P.$$

So consider $G \in \mathcal{A}_i$.
Now we will split G in subsets and prove the inequality above for each of these subsets.

Thus let us define for all $n \in \mathbb{N}_0$ with $i \leq n$

(5.11.14) $$A^{n,1} := G \cap \{\rho = n\} \cap \{\mathrm{E}\left(X_{\tau_{n+1}(C)}\big|\mathcal{A}_n\right) > X_n\}.$$

Let $n \in \mathbb{N}_0$ with $i \leq n$.

We have $A^{n,1} \in \mathcal{A}_i \subseteq \mathcal{A}_n$ and

(5.11.15)
$$\int_{A^{n,1}} X_\rho dP = \int_{A^{n,1}} X_n dP \leq \int_{A^{n,1}} \mathrm{E}\left(X_{\tau_{n+1}(C)}\big|\mathcal{A}_n\right) dP = \int_{A^{n,1}} X_{\tau_{n+1}(C)} dP = \int_{A^{n,1}} X_{\hat\rho} dP.$$

(5.11.13) shows that
$$G \cap \{\rho < \tau_\sigma(C^*)\} = \biguplus_{n=i}^{\infty} A^{n,1}.$$

Hence we have by (5.11.15) and by dominated convergence
$$\int_{G \cap \{\rho < \tau_\sigma(C^*)\}} X_\rho dP \leq \int_{G \cap \{\rho < \tau_\sigma(C^*)\}} X_{\hat\rho} dP.$$

We have by (5.11.12)
$$\rho = \hat\rho \text{ on } G \cap \{\rho = \tau_\sigma(C^*)\},$$

hence
$$\int_{G \cap \{\rho = \tau_\sigma(C^*)\}} X_\rho dP = \int_{G \cap \{\rho = \tau_\sigma(C^*)\}} X_{\hat\rho} dP.$$

So we have
$$\int_G X_\rho dP \leq \int_G X_{\hat\rho} dP. \qquad \square$$

Proof of (5.11.3) to (5.11.5) for arbitrary κ:
Obviously we have

(5.11.16) $\rho = \hat\rho = \tau_\sigma(C^*)$ on $\{\rho = \tau_\sigma(C^*)\} \stackrel{(5.11.2)}{\underset{(5.11.1)}{=}} \{\rho = \infty\} \cup \bigcup_{n=i}^{\infty} \{\rho = n\} \cap C_n^*$.

For all $n \in \mathbb{N}_0$ and for all $j \in \mathbb{N}$ with $j \leq \kappa$ we have

(5.11.17) on $\{\rho = n\} \cap \{j = \min\{l \in \mathbb{N}; X_n < \mathrm{E}\left(X_{\tau_{n+l}(C)}\big|\mathcal{A}_n\right)\}\}$
$$\rho < \rho + 1 = n + 1 \leq \underbrace{(\tau_{n+j}(C) \wedge \tau_\sigma(C^*))}_{=\hat\rho} \leq \tau_\sigma(C^*).$$

Furthermore we have

(5.11.18) $C_n \backslash C_n^* = \biguplus_{j=1}^{\kappa} \{j = \min\{l \in \mathbb{N}; X_n < \mathrm{E}\left(X_{\tau_{n+l}(C)}\big|\mathcal{A}_n\right)\}\}$ for all $n \in \mathbb{N}_0$, hence

$$
\begin{aligned}
(5.11.19) \quad & \{\rho < \tau_\sigma(C^*)\} \\
& \stackrel{(5.11.16)}{=} \biguplus_{n=i}^{\infty} \{\rho = n\} \setminus C_n^* \\
(5.11.20) \quad & \stackrel{(5.11.18)}{=} \biguplus_{n=i}^{\infty} \biguplus_{j=1}^{\kappa} \{\rho = n\} \cap \{j = \min\{l \in \mathbb{N};\ X_n < \mathrm{E}\left(X_{\tau_{n+l}(C)} | \mathcal{A}_n\right)\}\}. \quad \square
\end{aligned}
$$

Proof of (5.11.6) for arbitrary κ:
It suffices to prove for all $G \in \mathcal{A}_i$

$$\int_G X_\rho \mathrm{d}P \leqslant \int_G X_{\tilde\rho} \mathrm{d}P.$$

So consider $G \in \mathcal{A}_i$.
Now we will split G in subsets and prove the inequality above for each of these subsets.

Thus let us define for all $n \in \mathbb{N}_0$ with $i \leqslant n$, $j \in \mathbb{N}$ with $j \leqslant \kappa$

$$(5.11.21) \quad A^{n,j} := G \cap \{\rho = n\} \cap \{j = \min\{l \in \mathbb{N};\ X_n < \mathrm{E}\left(X_{\tau_{n+l}(C)} | \mathcal{A}_n\right)\}\}.$$

Let $n \in \mathbb{N}_0$ with $i \leqslant n$, $j \in \mathbb{N}$ with $j \leqslant \kappa$.

We have $G \in \mathcal{A}_i \subseteq \mathcal{A}_n$ and hence $A^{n,j} \in \mathcal{A}_n$.

Since

$$(5.11.22) \qquad A^{n,j} \stackrel{(5.11.21)}{\subseteq} \{X_n < \mathrm{E}\left(X_{\tau_{n+j}(C)} | \mathcal{A}_n\right)\}$$

we have

$$
\begin{aligned}
\int_{A^{n,j}} X_\rho \mathrm{d}P & \stackrel{(5.11.21)}{=} \int_{A^{n,j}} X_n \mathrm{d}P \\
& \stackrel{(5.11.22)}{\leqslant} \int_{A^{n,j}} \mathrm{E}\left(X_{\tau_{n+j}(C)} | \mathcal{A}_n\right) \mathrm{d}P \\
(5.11.23) \qquad & \stackrel{A^{n,j} \in \mathcal{A}_n}{=} \int_{A^{n,j}} X_{\tau_{n+j}(C)} \mathrm{d}P.
\end{aligned}
$$

Now we will define a disjoint decomposition of $A^{n,j}$. Thus we define

$$D := \{\tau_{n+j}(C) \leqslant \tau_\sigma(C^*)\}$$

and for all $m \in \mathbb{N}$ with $m < j$ define

$$A_m := A^{n,j} \cap \left(C_{n+m}^* \setminus \bigcup_{l=1}^{m-1} C_{n+l}^*\right).$$

Since $A^{n,j} \in \mathcal{A}_n$ we have

$$(5.11.24) \qquad A_m \in \mathcal{A}_{m+n} \text{ for all } m \in \mathbb{N} \text{ with } m < j.$$

We will prove now
$$A^{n,j} \cap D^c = \biguplus_{m=1}^{j-1} A_m.$$

Disjoint: For all $m, l \in \mathbb{N}$ with $m, l < j$, $m \neq l$ and (w.l.o.g.) $m \leq l$ we have $m \leq l - 1$ and thus
$$A_m \cap A_l \subseteq C^*_{n+m} \setminus C^*_{n+m} = \emptyset,$$
hence for all $m, l \in \mathbb{N}$ with $m, l < j$
$$m \neq l \implies A_m \cap A_l = \emptyset.$$

"$\supseteq$": For all $m \in \mathbb{N}$ with $m < j$ we have
$$A_m \subseteq D^c.$$

"$\subseteq$": Consider $\omega \in A^{n,j} \cap D^c$. Define
$$r := \tau_\sigma(C^*)(\omega) = \inf \{l \geq \sigma(\omega) \,;\, \omega \in C^*_l\}.$$
We have $r = \tau_\sigma(C^*)(\omega) \stackrel{(5.11.2)}{\geq} \rho(\omega) = n$.

Due to the definition of $A^{n,j}$ in (5.11.21) we have by (5.11.18) $\omega \in C_n \setminus C^*_n$, hence $r \neq n$.

Define
$$q := \tau_{n+j}(C)(\omega) = \inf \{l \geq n + j \,;\, \omega \in C_l\}.$$
For all $p \in \mathbb{N}_0$ with $n + j \leq p < q$ we have $\omega \notin C_l \supseteq C^*_l$.
Since $\omega \in D^c = \{\tau_{n+j}(C) > \tau_\sigma(C^*)\}$ we have $q > r$.
This yields $r < n + j$.

So we have $n < r < n + j$, thus $1 \leq r - n \leq j - 1$ and it follows $\omega \in A_{r-n}$ by definition of r.

By the above part of the proof we can deduce for all $m \in \mathbb{N}$ with $m < j$

(5.11.25) $$\tau_\sigma(C^*) = m + n \text{ on } A_m = A_{(m+n)-n}.$$

It follows

(5.11.26) $$A^{n,j} = \left(A^{n,j} \cap D\right) \uplus \biguplus_{m=1}^{j-1} A_m.$$

We have

(5.11.27) $$\tau_{n+j}(C) = \tau_{n+j}(C) \wedge \tau_\sigma(C^*) = \hat{\rho} \text{ on } A^{n,j} \cap D.$$

For all $m \in \mathbb{N}$ with $m < j$ we have

$$\int_{A_m} X_{\tau_{n+j}(C)} dP \overset{(5.11.24)}{=} \int_{A_m} \mathrm{E}\left(X_{\tau_{n+j}(C)} \big| \mathcal{A}_{n+m}\right) dP$$

$$\overset{A_m \subseteq C^*_{n+m}}{\underset{(5.9.1),\, j-m \leqslant \kappa}{\leqslant}} \int_{A_m} X_{n+m} dP$$

$$\overset{(5.11.25)}{=} \int_{A_m} X_{\tau_\sigma(C^*)} dP$$

(5.11.28) $$\overset{A_m \subseteq D^c}{=} \int_{A_m} X_{\hat{\rho}} dP.$$

By (5.11.26), (5.11.27) and (5.11.28) and summing over m we have

$$\int_{A^{n,j}} X_{\tau_{n+j}(C)} dP = \int_{A^{n,j}} X_{\hat{\rho}} dP.$$

Now we can combine this with (5.11.23) and so we have

$$\int_{A^{n,j}} X_\rho dP \overset{(5.11.23)}{\leqslant} \int_{A^{n,j}} X_{\tau_{n+j}(C)} dP$$

(5.11.29) $$= \int_{A^{n,j}} X_{\hat{\rho}} dP.$$

(5.11.20) (page 58) shows that

$$G \cap \{\rho < \tau_\sigma(C^*)\} = \biguplus_{n=i}^{\infty} \biguplus_{j=1}^{\kappa} A^{n,j}.$$

Hence we have by (5.11.29) and by dominated convergence

$$\int_{G \cap \{\rho < \tau_\sigma(C^*)\}} X_\rho dP \leqslant \int_{G \cap \{\rho < \tau_\sigma(C^*)\}} X_{\hat{\rho}} dP.$$

We have by (5.11.16) (page 57)

$$\rho = \hat{\rho} \text{ on } G \cap \{\rho = \tau_\sigma(C^*)\}.$$

hence

$$\int_{G \cap \{\rho = \tau_\sigma(C^*)\}} X_\rho dP = \int_{G \cap \{\rho = \tau_\sigma(C^*)\}} X_{\hat{\rho}} dP.$$

So we have

$$\int_G X_\rho dP \leqslant \int_G X_{\hat{\rho}} dP. \qquad \square$$

5.12 Theorem

$\tau^{(\infty)}$ is optimal in $\mathcal{S}(C^{(0)})$

and for all $j \in \mathbb{N}_0$, $m \in \mathbb{N}$ we have
$$\tau_j^{(m)} \leqslant \tau_j^{(m+1)},$$
$$Y_j^{(m)} \leqslant Y_j^{(m+1)}.$$

If we have $\tau^{(0)} \in \mathcal{S}(C^{(0)})$, this is also true for $m = 0$.

Proof: Consider $j \in \mathbb{N}_0$.

Step 1:
Consider $(C^{(m)})_{m \in \mathbb{N}_0}$ as constructed above.
Let $\rho \in \mathcal{S}(C^{(0)})$ with $\rho \geqslant j$.
Define $\rho' := \inf \left\{ n \; ; \; \rho \leqslant n, \mathbf{1}_{C_n^{(\infty)}} = 1 \right\}$, thus $\rho' \in \mathcal{S}(C^{(\infty)})$.
For all $m \in \mathbb{N}_0$ define $\rho_m := \inf \left\{ n \; ; \; \rho \leqslant n, \mathbf{1}_{C_n^{(m)}} = 1 \right\}$.
We have $\rho_0 = \inf \left\{ n \; ; \; \rho \leqslant n, \mathbf{1}_{C_n^{(0)}} = 1 \right\} = \inf \{ n \; ; \; \rho \leqslant n \} = \rho$,
because for all $n \in \mathbb{N}_0$ we have $\{\rho = n\} \subseteq C_n^{(0)}$.
Consider $m \in \mathbb{N}_0$.
For all $n \in \mathbb{N}_0 \cup \{\infty\}$ we have $C_n^{(m+1)} \subseteq C_n^{(m)}$ and thus

$$\left\{ \mathbf{1}_{C_n^{(m+1)}} = 1 \right\} \subseteq \left\{ \mathbf{1}_{C_n^{(m)}} = 1 \right\}.$$

So we have

$$\rho_{m+1} = \inf \left\{ n \; ; \; \rho \leqslant n, \mathbf{1}_{C_n^{(m+1)}} = 1 \right\} \geqslant \inf \left\{ n \; ; \; \rho \leqslant n, \mathbf{1}_{C_n^{(m)}} = 1 \right\} = \rho_m.$$

It follows by Lemma 5.11 (page 55), equation (5.11.7)
(adaption: $i = j$, $\sigma = \rho_m$, $\sigma^* = \rho_{m+1}$, $C = C^{(m)}$, $C^* = C^{(m+1)}$)

$$\mathrm{E}\left(X_{\rho_m} | \mathcal{A}_j\right) \leqslant \mathrm{E}\left(X_{\rho_{m+1}} | \mathcal{A}_j\right).$$

The facts $\rho_m \leqslant \rho_{m+1}$ for all $m \in \mathbb{N}_0$ and $\lim \rho_m = \rho'$ imply

$$\mathrm{E}\left(X_\rho | \mathcal{A}_j\right) \leqslant \lim_{m \to \infty} \mathrm{E}\left(X_{\rho_m} | \mathcal{A}_j\right) \stackrel{(5.2.2)}{=} \mathrm{E}\left(X_{\lim_{m \to \infty} \rho_m} \Big| \mathcal{A}_j\right) = \mathrm{E}\left(X_{\rho'} | \mathcal{A}_j\right).$$

So for every $\rho \in \mathcal{S}(C^{(0)})$ with $\rho \geqslant j$ there exists $\rho' \in \mathcal{S}(C^\infty)$ with $\mathrm{E}\left(X_\rho | \mathcal{A}_j\right) \leqslant \mathrm{E}\left(X_{\rho'} | \mathcal{A}_j\right)$.
For all $n \in \mathbb{N}_0 \cup \{\infty\}$ we have $C_n^{(\infty)} \subseteq C_n^{(0)}$ and thus $\mathcal{S}(C^{(\infty)}) \subseteq \mathcal{S}(C^{(0)})$.
The last two expressions yield:

$$\operatorname{esssup} \left\{ \mathrm{E}\left(X_\rho | \mathcal{A}_j\right) \; ; \; \rho \in \mathcal{S}(C^{(\infty)}), \rho \geqslant j \right\} = \operatorname{esssup} \left\{ \mathrm{E}\left(X_\rho | \mathcal{A}_j\right) \; ; \; \rho \in \mathcal{S}(C^{(0)}), \rho \geqslant j \right\}.$$

Step 2A:
Consider $\rho, \tau \in \mathcal{S}(C^{(\infty)})$ with $j \leqslant \rho \leqslant \tau$.
For all $m \in \mathbb{N}_0$ define

$$\rho_m := \tau \mathbf{1}_{\{\rho = \tau\}} + \sum_{n=j}^{\infty} \tau_{n+1}^{(m)} \mathbf{1}_{\{\rho = n\}} \mathbf{1}_{\{\rho < \tau\}}.$$

Define further
$$\widetilde{\rho} := \tau \mathbf{1}_{\{\rho=\tau\}} + \sum_{n=j}^{\infty} \tau_{n+1}^{(\infty)} \mathbf{1}_{\{\rho=n\}} \mathbf{1}_{\{\rho<\tau\}}.$$

We have $\rho \leqslant \rho_m \leqslant \rho_{m+1}$ for all $m \in \mathbb{N}_0$ and $\lim_{m\to\infty} \rho_m = \widetilde{\rho} \leqslant \tau$.

Consider $G \in \mathcal{A}_j$, $m \in \mathbb{N}_0$, $n \in \mathbb{N}_0$ with $j \leqslant n$.
We have
$$\begin{aligned} G \cap \{\rho = n\} \cap \{n < \tau\} &= G \cap \{\rho = n\} \cap \{\rho < \tau\} \\ &\subseteq \{\rho = n\} \\ &\stackrel{\rho \in \mathcal{S}(C^{(\infty)})}{\subseteq} C_n^{(\infty)} \\ &\stackrel{2.13}{\subseteq} C_n^{(m+1)} \\ &\stackrel{2.13}{\subseteq} \left\{ \mathrm{E}\left(X_{\tau_{n+1}^{(m)}} \middle| \mathcal{A}_n \right) \leqslant X_n \right\} \end{aligned}$$
and $G \cap \{\rho = n\} \cap \{n < \tau\} \in \mathcal{A}_n$, because $G \in \mathcal{A}_j \subseteq \mathcal{A}_n$ and $\{\rho = n\}, \{n < \tau\} \in \mathcal{A}_n$.
Thus
$$\int_{G\cap\{\rho=n\}\cap\{\rho<\tau\}} X_{\rho_m} dP = \int_{G\cap\{\rho=n\}\cap\{n<\tau\}} X_{\tau_{n+1}^{(m)}} dP \leqslant \int_{G\cap\{\rho=n\}\cap\{n<\tau\}} X_\rho dP.$$
So we have
$$\begin{aligned} \int_G X_{\rho_m} dP &= \int_{G\cap\{\rho=\tau\}} X_{\rho_m} dP + \sum_{n\in\mathbb{N}_0} \int_{G\cap\{\rho=n\}\cap\{\rho<\tau\}} X_{\rho_m} dP \\ &\leqslant \int_{G\cap\{\rho=\tau\}} X_\rho dP + \sum_{n\in\mathbb{N}_0} \int_{G\cap\{\rho=n\}\cap\{\rho<\tau\}} X_\rho dP \\ &= \int_G X_\rho dP. \end{aligned}$$

Combining these facts yields
$$\mathrm{E}\left(X_{\widetilde{\rho}}\middle|\mathcal{A}_j\right) = \mathrm{E}\left(X_{\lim_{m\to\infty}\rho_m}\middle|\mathcal{A}_j\right) \stackrel{(5.2.2)}{=} \lim_{m\to\infty} \mathrm{E}\left(X_{\rho_m}\middle|\mathcal{A}_j\right) \leqslant \mathrm{E}\left(X_\rho\middle|\mathcal{A}_j\right).$$

Step 2B:
Consider $\rho, \tau \in \mathcal{S}(C^\infty)$ with $j \leqslant \rho \leqslant \tau$.
Define $\rho_0 := \rho$ and for all $m \in \mathbb{N}_0$ define $\rho_{m+1} := \widetilde{\rho_m}$.
For all $m \in \mathbb{N}_0$ we have
$$\rho_m \leqslant \rho_{m+1} \leqslant \tau,$$
$$\rho_m + 1 \leqslant \widetilde{\rho_m} = \rho_{m+1} \text{ on } \{\rho_m < \tau\}.$$
It follows
$$\lim_{m\to\infty} \rho_m = \tau.$$

So for all $G \in \mathcal{A}_j$ we have

$$\mathrm{E}\left(X_\tau | \mathcal{A}_j\right) = \mathrm{E}\left(X_{\lim_{m \to \infty} \rho_m} \Big| \mathcal{A}_j\right) \stackrel{(5.2.2)}{=} \lim_{m \to \infty} \mathrm{E}\left(X_{\rho_m} | \mathcal{A}_j\right) \leqslant \mathrm{E}\left(X_\rho | \mathcal{A}_j\right).$$

Step 2C:
Let $\sigma \in \mathcal{S}(C^{(\infty)})$ with $\sigma \geqslant j$. Let $\omega \in \Omega$, $l := \sigma(\omega)$. Then $\omega \in C_l^{(\infty)}$ and $1 = \mathbf{1}_{C_l^{(\infty)}}(\omega)$, thus

$$\tau_j^{(\infty)}(\omega) = \inf\left\{ p \geqslant j \,;\, \mathbf{1}_{C_p^{(\infty)}}(\omega) = 1 \right\} \leqslant l = \sigma(\omega).$$

This yields $\tau_j^{(\infty)} \leqslant \sigma$.
So for all $\tau \in \mathcal{S}(C^{(\infty)})$ with $\tau \geqslant j$ we have $\tau \geqslant \tau_j^{(\infty)}$, thus

$$\mathrm{E}\left(X_\tau | \mathcal{A}_j\right) \leqslant \mathrm{E}\left(X_{\tau_j^{(\infty)}} \Big| \mathcal{A}_j\right).$$

This yields

$$\mathrm{E}\left(X_{\tau_j^{(\infty)}} \Big| \mathcal{A}_j\right) = \operatorname{esssup}\left\{ \mathrm{E}\left(X_\tau | \mathcal{A}_j\right) \,;\, \tau \in \mathcal{S}(C^{(\infty)}), \tau \geqslant j \right\}.$$

The above steps yield

$$\mathrm{E}\left(X_{\tau_j^{(\infty)}} \Big| \mathcal{A}_j\right) = \operatorname{esssup}\left\{ \mathrm{E}\left(X_\rho | \mathcal{A}_j\right) \,;\, \rho \in \mathcal{S}(C^{(0)}), \rho \geqslant j \right\}.$$

So for any family $C^{(0)} \in \mathcal{C}$ we have

$$\tau_j^{(\infty)} \in \mathcal{S}(C^{(0)}) \quad \text{for all } j \in \mathbb{N}_0. \qquad \square$$

5.13 Conclusion

We have for all $j \in \mathbb{N}_0$

$$Y_j^{(\infty)} = \mathrm{E}\left(X_{\tau_j^{(\infty)}} \Big| \mathcal{A}_j\right).$$

If there is some $k \in \mathbb{N}_0$ such that

$$\tau^{(k)} \text{ is optimal in } \mathcal{S}(C^{(k)})$$

or

$$C^{(k)} = C^{(k+1)}$$

then we have

$$\begin{aligned}
C^{(k)} &= C^{(k+n)} = C^{(\infty)} \text{ for all } n \in \mathbb{N}_0, \\
\tau^{(k)} &= \tau^{(k+n)} = \tau^{(\infty)} \text{ for all } n \in \mathbb{N}_0, \\
\Theta^{(k)} &= \Theta^{(k+n)} = \Theta^{(\infty)} \text{ for all } n \in \mathbb{N}_0, \\
Y^{(k)} &= Y^{(k+n)} = Y^{(\infty)} \text{ for all } n \in \mathbb{N}_0.
\end{aligned}$$

Proof: For all $j \in \mathbb{N}_0$ we have

$$Y_j^{(\infty)} \stackrel{\text{def.}}{\underset{\text{incr.}}{=}} \lim_{m \to \infty} Y_j^{(m)} \stackrel{\text{def.}}{=} \lim_{m \to \infty} \mathrm{E}\left(X_{\tau_j^{(m)}} \middle| \mathcal{A}_j\right)$$
$$\stackrel{\text{ass.}}{=} \mathrm{E}\left(X_{\left(\lim_{m \to \infty} \tau_j^{(m)}\right)} \middle| \mathcal{A}_j\right) \stackrel{\text{def.}}{\underset{\text{incr.}}{=}} \mathrm{E}\left(X_{\tau_j^{(\infty)}} \middle| \mathcal{A}_j\right).$$

The other statement is true by Lemma 5.10 (page 55) and the definitions in Part 2.13 (page 22). □

5.14 Conclusion

If $C^{(0)}$ is essential, then
$$\tau^{(\infty)} \text{ is optimal.}$$

Proof: Assume $C^{(0)}$ is essential. Then there is some optimal $\sigma \in \mathcal{T}$ with $\sigma_n \in \mathcal{S}(C^{(0)})$ for all $n \in \mathbb{N}_0$. So we have for all $n \in \mathbb{N}_0$

$$\mathrm{E}(X_{\sigma_n}|\mathcal{A}_n) \leqslant \operatorname{esssup}\{\mathrm{E}(X_\rho|\mathcal{A}_n) \ ; \ \rho \in \mathcal{S}_n^\infty \cap \mathcal{S}(C^{(0)})\}$$
$$\leqslant \operatorname{esssup}\{\mathrm{E}(X_\rho|\mathcal{A}_n) \ ; \ \rho \in \mathcal{S}_n^\infty\}$$
$$= \mathrm{E}(X_{\sigma_n}|\mathcal{A}_n).$$

By Theorem 5.12 (page 61)
$$\tau^{(\infty)} \text{ is optimal in } \mathcal{S}(C^{(0)}),$$

hence for all $n \in \mathbb{N}_0$ we have
$$\mathrm{E}\left(X_{\tau_n^{(\infty)}} \middle| \mathcal{A}_n\right) = \operatorname{esssup}\{\mathrm{E}(X_\rho|\mathcal{A}_n) \ ; \ \rho \in \mathcal{S}_n^\infty \cap \mathcal{S}(C^{(0)})\}.$$

Hence we have for all $n \in \mathbb{N}_0$
$$\mathrm{E}\left(X_{\tau_n^{(\infty)}} \middle| \mathcal{A}_n\right) = \mathrm{E}(X_{\sigma_n}|\mathcal{A}_n),$$

and
$$\tau^{(\infty)} \text{ is optimal.}$$ □

5.15 Remark

If $C^{(0)} \equiv \Omega$, then
$$\tau^{(\infty)} \text{ is optimal.}$$

Proof: Consider $C^{(0)} \equiv \Omega$. There are two ways to prove the optimality:
(1) We have $\mathcal{S} = \mathcal{S}(C^{(0)})$ and hence for all $j \in \mathbb{N}_0$ we have
$$\mathrm{E}\left(X_{\tau_j^{(\infty)}} \middle| \mathcal{A}_j\right) = \operatorname{esssup}\{\mathrm{E}(X_\rho|\mathcal{A}_j) \ ; \ \rho \in \mathcal{S}, \rho \geqslant j\}.$$
(2) $C^{(0)}$ is essential, hence $\tau^{(\infty)}$ is optimal by Conclusion 5.14 (page 64). □

5.16 Outlook

The above theorems and proofs are adapted to the Markovian case with random discounting in Section 7 (page 71). The Markovian case itself is introduced in the following Section 6 (page 65).

6 Markovian Case

In this chapter we adapt the FII Algorithm stated in Section 2 (page 15) to Markovian stopping problems.

After repeating the setting in Part 6.1 (page 65), we consider the case of time-independent pay-off for the case of arbitrary κ in Subsection 6.2 (page 65). For $\kappa \equiv 1$ this was done before in [Irl06, Part 2] and [Irl09, Part 1]. Some cases of time-dependent pay-off for $\kappa \equiv 1$ have been mentioned in [Irl09, p. 2] and in detail considered in [Irl06, part 3.1]. Here we will show their extensions to arbitrary κ in Subsection 6.3 (page 67). Part 6.4 (page 68) shows that the case of certain path-dependent pay-offs, considered in [Pre10], fits in our setting as well.

6.1 Setting

Let $(Z_n)_{n \in \mathbb{N}_0}$ be a time-homogeneous Markov process with respect to the underlying filtration with state space $(S, \mathcal{S})$. In the Markovian case the FII algorithm only has to work in the systems of subsets of the state space S, as we will show exemplary for the case of time-independent pay-off in Part 6.2.1 (page 66).

6.2 Time-Independent Pay-Off

For all $B \in \mathcal{S}$ define
$$\mathcal{T}(B) := \{\tau \in \mathcal{T} \; ; \; Z_\tau \in B \text{ on } \{\tau < \infty\}\},$$
$$\tau_n(B) := \inf\{k \geq n : Z_k \in B\} \text{ for all } n \in \mathbb{N}_0.$$

Given a measurable mapping $g : S \to \mathbb{R}$ consider the problem of optimal stopping for the pay-off
$$X_n := g(Z_n) \text{ for all } n \in \mathbb{N}_0,$$
defining for simpler notation
$$X_\infty := g(Z_\infty) := \lim_{n \to \infty} \sup_{k \geq n} g(Z_k),$$
so that we have
$$X_\tau = g(Z_\tau) \text{ for all } \tau \in \mathcal{T}.$$

Assume as usual
$$\mathrm{E}(X_\tau | Z_0 = z) \text{ exists for all } \tau \in \mathcal{T} \text{ and } z \in S.$$

Then the iterative step is performed for given κ by
$$B^{*\kappa} := \left\{ z \in B \; ; \; g(z) \geq \sup_{1 \leq j < \kappa+1} \mathrm{E}\left(g\left(Z_{\tau_j(B)}\right) \middle| Z_0 = z\right) \right\} \text{ for all } B \in \mathcal{S}.$$

6.2.1 Connection to the General Case

Consider N infinite and some $n \in \mathbb{N}_0$ with $C^{(n)}$ given such that there is some B with

$$C_k^{(n)} = \{Z_k \in B\} \text{ for all } k \in \mathbb{N}_0.$$

Then we have

$$C_k^{(n+1)} = \{Z_k \in B^{*\kappa(n)}\} \text{ for all } k \in \mathbb{N}_0.$$

Therefore the FII algorithm only has to work in the systems of subsets of the state space S.

Proof: We have $\tau_k^{(n)} = \inf\{p \geq k \; ; \; Z_p \in B\} = \tau_k(B)$ for all $k \in \mathbb{N}_0$.
Assume $(Z'_n)_{n \in \mathbb{N}_0}$ being an independent copy of $(Z_n)_{n \in \mathbb{N}_0}$ and

$$\tau'^{(n)}_p = \tau'_p(B) = \inf\{q \geq p \; ; \; Z'_q \in B\} \text{ for all } p \in \mathbb{N}_0.$$

We have for all $p, k \in \mathbb{N}_0$ with $p \leq k$

$$\left\{ \mathrm{E}\left(g(Z_{\tau_p^{(n)}}) \big| \mathcal{A}_k\right) \leq g(Z_k) \right\}$$
$$= \left\{ Z_k \in \left\{ \mathrm{E}\left(g(Z'_{\tau'^{(n)}_p}) \big| Z'_k = z\right) \leq g(z) \right\} \right\}$$
$$= \left\{ Z_k \in \left\{ \mathrm{E}\left(g(Z'_{\tau'^{(n)}_{p-k}}) \big| Z'_0 = z\right) \leq g(z) \right\} \right\}$$
$$= \left\{ Z_k \in \left\{ \mathrm{E}\left(g(Z'_{\tau'_{p-k}(B)}) \big| Z'_0 = z\right) \leq g(z) \right\} \right\}.$$

Hence we have

$$C_k^{(n+1)} = C_k^{(n)} \cap \bigcap \left\{ \left\{ \mathrm{E}\left(X_{\tau_p^{(n)}} \big| \mathcal{A}_k\right) \leq X_k \right\} \; ; \; k \leq p \leq k + \kappa(n) \right\}$$
$$= \{Z_k \in B\} \cap \bigcap \left\{ Z_k \in \left\{ g(z) \geq \mathrm{E}\left(g(Z'_{\tau'_p(B)}) \big| Z'_0 = z\right) \right\} \; ; \; 0 \leq p \leq \kappa(n) \right\}$$
$$= \{Z_k \in B\} \cap \left\{ Z_k \in \left\{ g(z) \geq \sup \left\{ \mathrm{E}\left(g(Z'_{\tau'_p(B)}) \big| Z'_0 = z\right) \; ; \; 0 \leq p \leq \kappa(n) \right\} \right\} \right\}$$
$$= \left\{ Z_k \in \left\{ z \in B \; ; \; g(z) \geq \sup \left\{ \mathrm{E}\left(g(Z'_{\tau'_p(B)}) \big| Z'_0 = z\right) \; ; \; 0 \leq p \leq \kappa(n) \right\} \right\} \right\}$$
$$= \{Z_k \in B^{*\kappa(n)}\}. \qquad \square$$

6.3 Time-Dependent Pay-Off

We can apply the algorithm and its results to the "time-independent" space-time Markov process $(Z_n, n)_{n \in \mathbb{N}_0}$ on $S \times \mathbb{N}_0$ instead of S. Hence the time-dependent case can be treated as follows:

Consider $f : S \times \mathbb{N}_0 \to \mathbb{R}$ and the problem of optimal stopping for the pay-off

$$X_n := f(Z_n, n) \text{ for all } n \in \mathbb{N}_0,$$

defining (as above) for simpler notation

$$X_\infty := f(Z_\infty, \infty) := \limsup_{n \to \infty \, k \geq n} f(Z_k, k),$$

so that we have

$$X_\tau = f(Z_\tau, \tau) \text{ for all stopping rules } \tau.$$

Assume as usual

$$\mathrm{E}\left(X_\tau | Z_0 = z\right) \text{ exists for all } \tau \in \mathcal{T} \text{ and } z \in S.$$

Then the iterative step is done for any κ and any measurable $B \subseteq S \times \mathbb{N}_0$ by

$$B^{*\kappa} := \left\{ (z, n) \in B \ ; \ f(z, n) \geq \sup_{1 \leq j < \kappa + 1} \mathrm{E}\left(f\left(Z_{\tau_j(B)}, n + \tau_j(B)\right) \middle| Z_0 = z\right) \right\}$$

with

$$\tau_n(B) := \inf \{ k \geq n \ ; \ (Z_k, k) \in B \} \text{ for all } n \in \mathbb{N}_0.$$

6.3.1 Finite Horizon

In the case of a stopping problem with finite time horizon $N \in \mathbb{N}$ and pay-off

$$f(Z_n, n) \text{ for all } n \in \mathbb{N}_0 \text{ with } n \leq N$$

just define

$$f(z, k) = -\infty \text{ for all } z \in S \text{ and for all } k \in \mathbb{N}_0 \text{ with } N + 1 \leq k.$$

Then for any κ and any measurable $B \subseteq S \times \mathbb{N}_0$ we have $B^{*\kappa} \subseteq S \times \{0, ..., N\}$.

6.3.2 Linear Costs

Consider some measurable mapping $g : S \to \mathbb{R}$, some $c \in \mathbb{R}_{>0}$ (often representing the costs of observation for one time step) and

$$f(z, n) = g(z) - cn \text{ for all } z \in S \text{ and for all } n \in \mathbb{N}_0.$$

Then for all $z \in S$, $n \in \mathbb{N}_0$ and $j \in \mathbb{N}$ we have

$$\begin{aligned}
& \mathrm{E}\left(f\left(Z_{\tau_j(B)}, n + \tau_j(B)\right) \middle| Z_0 = z\right) \\
={} & \mathrm{E}\left(g\left(Z_{\tau_j(B)}\right) - cn - c\tau_j(B) \middle| Z_0 = z\right) \\
={} & \mathrm{E}\left(g\left(Z_{\tau_j(B)}\right) \middle| Z_0 = z\right) - cn - c\mathrm{E}\left(\tau_j(B) \middle| Z_0 = z\right),
\end{aligned}$$

hence the iterative step is done for any κ and any measurable $B \subseteq S \times \mathbb{N}_0$ by

$$\begin{aligned} B^{**\kappa} &:= \left\{ (z,n) \in B \,;\, f(z,n) \geq \sup_{1 \leq j \leq \kappa+1} \mathrm{E}\left(f\left(Z_{\tau_j(B)}, n + \tau_j(B)\right) \big| Z_0 = z\right) \right\} \\ &= \left\{ (z,n) \in B \,;\, g(z) \geq \sup_{1 \leq j \leq \kappa+1} \left(\mathrm{E}\left(g(Z_{\tau_j(B)}) \big| Z_0 = z\right) - c\mathrm{E}\left(\tau_j(B) | Z_0 = z\right) \right) \right\}. \end{aligned}$$

Obviously there is no time-dependence any more.
This example is extended in Part 6.4 (page 68).

6.3.3 Constant Discounting

Consider some measurable mapping $g : S \to \mathbb{R}$, some (constant discount factor) $\alpha \in]0,1]$ and

$$f(z,n) = \alpha^n g(z) \text{ for all } z \in S \text{ and for all } n \in \mathbb{N}_0$$

Then for all $z \in S$, $n \in \mathbb{N}_0$ and $j \in \mathbb{N}$ we have

$$\begin{aligned} & \mathrm{E}\left(f\left(Z_{\tau_j(B)}, n + \tau_j(B)\right) \big| Z_0 = z\right) \\ &= \mathrm{E}\left(\alpha^{n+\tau_j(B)} g\left(Z_{\tau_j(B)}\right) \big| Z_0 = z\right) \\ &= \alpha^n \mathrm{E}\left(\alpha^{\tau_j(B)} g\left(Z_{\tau_j(B)}\right) \big| Z_0 = z\right), \end{aligned}$$

hence the iterative step is done for any κ and any measurable $B \subseteq S \times \mathbb{N}_0$ by

$$\begin{aligned} B^{**\kappa} &:= \left\{ (z,n) \in B \,;\, f(z,n) \geq \sup_{1 \leq j \leq \kappa+1} \mathrm{E}\left(f\left(Z_{\tau_j(B)}, n + \tau_j(B)\right) \big| Z_0 = z\right) \right\} \\ &= \left\{ (z,n) \in B \,;\, g(z) \geq \sup_{1 \leq j \leq \kappa+1} \mathrm{E}\left(\alpha^{\tau_j(B)} g\left(Z_{\tau_j(B)}\right) \big| Z_0 = z\right) \right\}. \end{aligned}$$

Again there is no time-dependence any more.
We consider the case of random discounting in detail in Section 7 (page 71).

6.4 Path-Dependent Costs

Consider some measurable mappings $g, c : S \to \mathbb{R}$. $g(z)$ is the payoff for stopping the observation and $c(z)$ is the cost of continuation for each point z. The simplification of constant c boils down to the case of linear costs considered in Part 6.3.2 (page 67). Consider the pay-off

$$X_n := g(Z_n) - \sum_{i=0}^{n-1} c(Z_i) \text{ for all } n \in \mathbb{N}_0.$$

The payoff is path-dependent. Now we apply the algorithm and its results to a time-independent "space-cost" Markov process $(Z_n, \sum_{i=0}^{n-1} c(Z_i))$ on $S \times \mathbb{R}$ instead of S. So we can treat the problem as follows:
Consider

$$f : S \times \mathbb{R} \longrightarrow \mathbb{R}, \; (z, d) \longmapsto g(z) - d$$

and the problem of optimal stopping for the pay-off

$$X_n := f\left(Z_n, \sum_{i=0}^{n-1} c(Z_i)\right) = g(Z_n) - \sum_{i=0}^{n-1} c(Z_i) \text{ for all } n \in \mathbb{N}_0,$$

defining (as above) for simpler notation

$$X_\infty := f\left(Z_\infty, \sum_{i=0}^{\infty-1} c(Z_i)\right) := \limsup_{n\to\infty \atop k\geqslant n} f\left(Z_k, \sum_{i=0}^{k-1} c(Z_i)\right),$$

so that we have

$$X_\tau = \left(Z_\tau, \sum_{i=0}^{\tau-1} c(Z_i)\right) \text{ for all stopping rules } \tau.$$

Assume as usual

$$\mathrm{E}\left(X_\tau | Z_0 = z\right) \text{ exists for all } \tau \in \mathcal{T} \text{ and } z \in S.$$

Then the iterative step is done for any κ and any measurable $B \subseteq S \times \mathbb{R}$ by

$$B^{*\kappa} := \left\{(z,d) \in B \,;\, f(z,d) \geqslant \sup_{1\leqslant j < \kappa+1} \mathrm{E}\left(f\left(Z_{\tau_j(B)}, \sum_{i=0}^{\tau_j(B)-1} c(Z_i) + d\right) \Big| Z_0 = z\right)\right\}$$

$$= \left\{(z,d) \in B \,;\, g(z) \geqslant \sup_{1\leqslant j < \kappa+1} \mathrm{E}\left(g(Z_{\tau_j(B)}) - \sum_{i=0}^{\tau_j(B)-1} c(Z_i) \Big| Z_0 = z\right)\right\}.$$

Obviously there is no path-dependence any more.

This problem has been considered by Presman in [Pre10]. The iteration is therein only explicitly defined under certain additional assumptions for B of the form $B = S \backslash C_k$ in Lemma 1 as follows

$$\begin{aligned}
D_{k+1} = S \backslash C_{k+1} &= S \backslash (C_k \cup \{z \in S \,;\, g_k(z) < Tg_k(z)\}) \\
&= S \backslash (C_k \cup \{z \in S \,;\, g_{C_k}(z) < Tg_{C_k}(z)\}) \\
&= (S \backslash C_k) \cap (S \backslash \{z \in S \,;\, g_{C_k}(z) < Tg_{C_k}(z)\}) \\
&= \{z \in S \backslash C_k \,;\, g_{C_k}(z) < Tg_{C_k}(z)\}
\end{aligned}$$

This can be rewritten in the form of our algorithm above by checking that

$$B^{*1} = \{z \in B \,;\, g_B(z) < Tg_B(z)\}$$

for each B and the appropriately defined functions g_B and T in [Pre10].

6.5 Performing the Algorithmic Step

In [Irl06, part 4] the algorithmic step of going from B to B^{*1} was performed by providing numerical values for the conditional expectations by *path-wise simulations of the Markov chain*. In the examples of [Irl06, part 4] the state space was finite and the algorithm found the solution within a surprisingly low number of iterations.

Rewriting a problem with finite state space as a *linear equation* as done in [Irl09, part 2] is another approach. It is therein done for $\kappa \equiv 1$. We will show this here for arbitrary κ in Section 8 (page 85).

6.6 Outlook

The theorems and proofs given in Section 5 (page 51) for the general case will be adapted to the Markovian case with random discounting in Section 7 (page 71). The same situation under finite state space is considered in Section 8 (page 85): Finding the solution can be rewritten as solving linear equations. Numerical examples calculated this way will be given in Section 9 (page 89).

7 Markovian Case with Random Discounting

The proofs given in Section 5 (page 51) for the general case will be adapted to the Markovian case with random discounting in this section. Herein we will also give some examples to explain the origin and to confirm the necessity of some definitions within the algorithm. Simplifications of the calculations for the case of finite state space will be examined in Section 8 (page 85). Numerical examples can be found in Section 9 (page 89).

We motivate in Part 7.1 (page 71) the setting we will give in Setting 7.2 (page 72). The main theorem of this chapter is Theorem 7.11 (page 81). This theorem is based on Lemma 7.4 (page 73), which is proved in Proof 7.9 (page 78) and uses therein Lemma 7.7 (page 76) and Lemma 7.6 (page 75). Example 7.5 (page 74) and Example 7.8 (page 76) substantiate necessities in these lemmata.

Lemma 7.4 (page 73) is adapted from [Irl09, Part 3, Theorem 1, Proof, Part (a)], where the case of $D = \{1\}$ is considered, now proved for almost arbitrary D. Theorem 7.11 (page 81) is based on ideas of [Irl09, Part 3, Theorem 1], where $\kappa = 1$ is used. The basic ideas are from [Irl80].

7.1 Motivation

Consider a (time-homogeneous) Markov chain in continuous time $(\widehat{Z}_t)_{t \in [0,\infty[}$ operating on the discrete state space $(S, \mathcal{P}ot(S))$ and a payoff

$$\widehat{X}_t := \widehat{\alpha}^t g(\widehat{Z}_t) \text{ for each } t \in [0, \infty[\text{ and } \widehat{X}_\infty := \lim_{t \to \infty} \sup_{s \geqslant t} \widehat{X}_s.$$

Assuming no instantaneous jumps (i.e. between two jump times there is always a positive amount of time) and no absorption, the above Markov chain has a pure jump structure, it is a special Markov jump process and we can describe it up to explosion time completely by a sequence of jump times and a sequence of states visited with adequate parameters, see [Asm03, p. 39-40]. When focussing on the problem of optimal stopping of the above Markov chain, we can reduce this problem to one of the following setting as described in [Irl09, part 3].

7.2 Setting

Let $(Z_n)_{n\in\mathbb{N}_0}$ be a time-homogeneous Markov process with respect to the underlying filtration and the discrete state space $(S, \mathcal{P}ot(S))$, $g: S \longrightarrow \mathbb{R}$, $\alpha: S \longrightarrow [0,1]$.
We look at the problem of optimal stopping for the pay-off

$$X_n := \left(\prod_{i=0}^{n-1} \alpha(Z_i)\right) g(Z_n) \text{ for all } n \in \mathbb{N}_0, \quad X_\infty := \limsup_{n\to\infty} \sup_{k \geq n} X_k.$$

We write for simpler notation formally

$$X_\infty = \left(\prod_{i=0}^{\infty} \alpha(Z_i)\right) g(Z_\infty)$$

so that we have

$$X_\tau = \left(\prod_{i=0}^{\tau-1} \alpha(Z_i)\right) g(Z_\tau) \text{ for all stopping rules } \tau.$$

Assume as usual

(7.2.1) $\qquad\qquad\qquad \mathbb{E}(X_\tau|Z_0 = z)$ exists for all $\tau \in \mathcal{T}$ and $z \in S$.

For all $B \in \mathcal{S}$ define

$$\mathcal{T}(B) := \{\tau \in \mathcal{T} \; ; \; Z_\tau \in B \text{ on } \{\tau < \infty\}\}.$$

For all $B \in \mathcal{S}$ and $\sigma \in \mathcal{T}$ define

(7.2.2) $\qquad\qquad\qquad \tau_\sigma(B) := \inf\{k \geq \sigma \; ; \; Z_k \in B\},$

the time of first entrance in B at or after time σ,
and if $\sigma \equiv n$ for some $n \in \mathbb{N}_0$ write $\tau_n(B)$ for $\tau_\sigma(B)$.

Define for all $B \in \mathcal{S}$ and $k \in \mathbb{N}$

$$B^{*k} := \{z \in B \; ; \; g(z) \geq \sup\{\mathbb{E}(X_{\tau_l(B)}|Z_0 = z) \; ; \; l \in \mathbb{N}_{\leq k}\}\}.$$

While running numerical examples (see Chapter 9 (page 89)) it appeared that using any $D \subseteq \mathbb{N}$ with $1 \in D$ for the improvement step instead of an initial sequence of $\mathbb{N}$ gives also optimal results and the convergence speed depends only on the cardinality of D not on the largest element of D. Hence let us define for all $B \in \mathcal{S}$ and $D \subseteq \mathbb{N}$

$$B^{*D} := \{z \in B \; ; \; g(z) \geq \sup\{\mathbb{E}(X_{\tau_k(B)}|Z_0 = z) \; ; \; k \in D\}\}.$$

This extends the definition above since $B^{*k} = B^{*\mathbb{N}_{\leq k}}$.

We tried to prove all results for arbitrary D — and for the first part of Lemma 7.4 (page 73) this worked. But already for the second part we had to draw back to an initial sequence of $\mathbb{N}$. The reason is shown in Example 7.8 (page 76).

For all $B \in \mathcal{S}$, $D \subseteq \mathbb{N}$, $\sigma \in \mathcal{T}(B)$ and $\rho \in \mathcal{T}(B)$ with $\sigma \leq \rho \leq \tau_\sigma(B^{*D})$ define

(7.2.3) $\qquad \hat{\rho} := \rho \mathbf{1}_{\{\rho = \tau_\sigma(B^{*D})\}} + \sum_{n=0}^{\infty}\sum_{j \in D} \mathbf{1}_{\{\rho=n\}\cap\{j=\inf\{i\in D \; ; \; Z_n \notin B^{*D\leq i}\}\}} \left(\tau_{n+j}(B) \wedge \tau_\sigma(B^{*D})\right).$

7.3 Remarks

- The time-independent case is included by setting $\alpha \equiv 1$.
- The situation of constant discounting as described in Part 6.3.3 (page 68) is included by assuming the function α being constant.
- If $D = \{1\}$, then we have by (7.2.3)

$$\hat{\rho} = \rho \mathbf{1}_{\{\rho = \tau_\sigma(B^{*\{1\}})\}} + \sum_{n=0}^{\infty} \mathbf{1}_{\{\rho = n\}} \tau_{n+1}(B).$$

7.4 Lemma: One-Step Improvement

Let $B \in \mathcal{S}$, $D \subseteq \mathbb{N}$, $\sigma \in \mathcal{T}(B)$ and

(7.4.1) $\qquad \rho \in \mathcal{T}(B)$ with $\sigma \leqslant \rho \leqslant \tau_\sigma(B^{*D})$.

Then we have

(7.4.2) $\qquad \hat{\rho} \in \mathcal{T}(B)$ with $\sigma \leqslant \hat{\rho} \leqslant \tau_\sigma(B^{*D})$,
(7.4.3) $\qquad \rho \leqslant \hat{\rho}$,
(7.4.4) $\qquad \rho + 1 \leqslant \hat{\rho}$ on $\{\rho < \tau_\sigma(B^{*D})\}$.

Assume additionally the existence of $k \in \mathbb{N} \cup \{\infty\}$ with $D = \mathbb{N}_{\leqslant k}$. Then we have

(7.4.5) $\qquad \mathrm{E}\left(X_\rho | Z_0 = z\right) \leqslant \mathrm{E}\left(X_{\hat{\rho}} | Z_0 = z\right)$ for all $z \in S$,

and if additionally

(7.4.6) $\qquad \mathrm{E}\left(X_{\lim_{n \to \infty} \sigma_n} \middle| Z_0 = z\right) = \lim_{n \to \infty} \mathrm{E}\left(X_{\sigma_n} | Z_0 = z\right)$
$\qquad$ for all $z \in S$ and all non-decreasing sequences $(\sigma_n)_{n \in \mathbb{N}_0}$ of stopping rules,

then we have

(7.4.7) $\qquad \mathrm{E}\left(X_\sigma | Z_0 = z\right) \leqslant \mathrm{E}\left(X_{\tau_\sigma(B^{*D})} \middle| Z_0 = z\right)$ for all $z \in S$.

Remark:

- This lemma is an adaption of Lemma 5.11 (page 55) above to Markovian stopping problems.
- The proof is postponed, see Proof 7.9 (page 78).

7.5 Example: Insufficient Improvement

Using the definition

$$\hat{\rho} := \rho \mathbb{1}_{\{\rho = \tau_\sigma(B^{*D})\}} + \sum_{n=0}^{\infty} \sum_{j \in D} \mathbb{1}_{\{\rho=n\} \cap \{j = \inf\{i \in D \,;\, Z_n \notin B^{*D \leq i}\}\}} \tau_{n+j}(B)$$

instead of (7.2.3) may contradict (7.4.2) and (7.4.7).

We can see this by the following example:
Imagine a Markov chain with five states a, b, c, d, e and the following transition matrix

	a	b	c	d	e
a	0	$\frac{1}{3}$	$\frac{1}{3}$	$\frac{1}{3}$	0
b	0	1	0	0	0
c	0	1	0	0	0
d	0	0	0	0	1
e	0	0	0	0	1

as drawn in the following picture

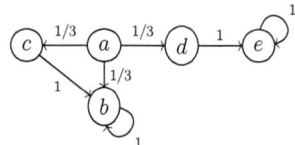

and the payments, i.e. the function g

state	payment
a	3
b	4
c	1.5
d	2.5
e	2

and we will use $D := \{1, 2\}$, $\alpha := 1$. We initialize the algorithm with

$$B = \{a, b, c, d, e\}.$$

Hence we have

$$B^{*\{1\}} = \{a, b, d, e\}$$

and

$$B^{*\{1,2\}} = \{b, d, e\}.$$

Now we determine $\hat{\rho}$ for $\rho := \tau_\sigma(B^{*\{1\}})$. On $\{Z_0 \neq a\}$ we have $\hat{\rho} = \tau_\sigma(B^{*\{1,2\}})$.
If $Z_0 = a$, then $\rho = n = 0$, hence

$$j := \inf \{i \in D \,;\, Z_n \notin B^{*D \leq i}\} = \inf \{i \in D \,;\, a \notin B^{*D \leq i}\} = 2$$

and thus on $\{Z_0 = a\}$

$$\hat{\rho} = \tau_{n+j}(B) = \tau_{0+2}(B) = \tau_2(B) = \inf\{n \geqslant 2 \,;\, Z_n \in \{b, e\}\}$$
$$> \inf\{n \geqslant 0 \,;\, Z_n \in \{b, d, e\}\} = \tau_\sigma(B^{*\{1,2\}}).$$

This is contradicting (7.4.2), and this $\hat{\rho}$ gives a smaller expected revenue, contradicting (7.4.7).

Hence it is necessary to take the minimum with σ^{*D} for a sufficient improvement, in the sense of having $\hat{\rho}$ always smaller than σ^{*D} and giving not a smaller expected revenue.

7.6 Lemma: Partial Improving I

Assume $B \in \mathcal{S}$, $D \subseteq \mathbb{N}$, $n \in \mathbb{N}_0$, $j \in D$, $A \in \mathcal{A}_n$ with $A \subseteq \{Z_n \in B^{*D<j} \setminus B^{*D\leqslant j}\}$, $z \in S$. Then we have

$$\int_A X_n \mathrm{d}P_z \leqslant \int_A X_{\tau_{n+j}(B)} \mathrm{d}P_z.$$

Proof: For all $E \subseteq D$, $j \in E$ we have

$$B^{*E} \subseteq B^{*E \setminus \{j\}},$$
$$B^{*E} = \{z \in B^{*E \setminus \{j\}} \,;\, g(z) \geqslant \mathrm{E}\left(X_{\tau_j(B)} \middle| Z_0 = z\right)\},$$
(7.6.1) $\quad B^{*E \setminus \{j\}} \setminus B^{*E} = \{z \in B^{*E \setminus \{j\}} \,;\, g(z) < \mathrm{E}\left(X_{\tau_j(B)} \middle| Z_0 = z\right)\}.$

Assume $(Z'_m)_{m \in \mathbb{N}_0}$ being an independent copy of $(Z_m)_{m \in \mathbb{N}_0}$ and

$$\tau'_p(B) = \inf\{q \geqslant p \,;\, Z'_q \in B\} \text{ for all } p \in \mathbb{N}_0.$$

Then we have

$$\int_A X_n \mathrm{d}P_z$$

$$\overset{\text{def. of } X_n}{=} \int_A \left(\prod_{i=0}^{n-1} \alpha(Z_i)\right) g(Z_n) \mathrm{d}P_z$$

$$\overset{\substack{Z_n \in B^{*D<j} \setminus B^{*D\leqslant j} \\ \leqslant \\ (7.6.1)}}{\leqslant} \int_A \left(\prod_{i=0}^{n-1} \alpha(Z_i)\right) \mathrm{E}\left(g\left(Z'_{\tau'_j(B)}\right) \prod_{i=0}^{\tau'_j(B)-1} \alpha(Z'_i) \middle| Z'_0 = Z_n\right) \mathrm{d}P_z$$

$$\overset{\text{M.C.}}{=} \int_A \left(\prod_{i=0}^{n-1} \alpha(Z_i)\right) \mathrm{E}\left(g\left(Z_{\tau_{n+j}(B)}\right) \prod_{i=n}^{\tau_{n+j}(B)-1} \alpha(Z_i) \middle| \mathcal{A}_n\right) \mathrm{d}P_z$$

$$\overset{A \in \mathcal{A}_n}{=} \int_A \left(\prod_{i=0}^{n-1} \alpha(Z_i)\right) g\left(Z_{\tau_{n+j}(B)}\right) \prod_{i=n}^{\tau_{n+j}(B)-1} \alpha(Z_i) \mathrm{d}P_z$$

$$= \int_A \left(\prod_{i=0}^{\tau_{n+j}(B)-1} \alpha(Z_i)\right) g\left(Z_{\tau_{n+j}(B)}\right) \mathrm{d}P_z$$

$$\overset{\text{def.}}{=} \int_A X_{\tau_{n+j}(B)} \mathrm{d}P_z.$$

$\square$

7.7 Lemma: Partial Improving II

Assume $B \in \mathcal{S}$, $D \subseteq \mathbb{N}$, $s,t \in \mathbb{N}$ with $s \leqslant t$ and $t - s \in D$, $A \in \mathcal{A}_s$ with $A \subseteq \{Z_s \in B^{*D}\}$. Then we have

(7.7.1) $$\int_A X_{\tau_t(B)} \mathrm{d}P_z \leqslant \int_A X_s \mathrm{d}P_z.$$

Proof: Assume $(Z'_n)_{n \in \mathbb{N}_0}$ being an independent copy of $(Z_n)_{n \in \mathbb{N}_0}$ and
$$\tau'_p(B) = \inf \{q \geqslant p \; ; \; Z'_q \in B\} \text{ for all } p \in \mathbb{N}_0.$$

Then we have

$$\int_A X_{\tau_t(B)} \mathrm{d}P_z$$

$$\stackrel{\text{def.}}{=} \int_A \left(\prod_{i=0}^{\tau_t(B)-1} \alpha(Z_i)\right) g\left(Z_{\tau_t(B)}\right) \mathrm{d}P_z$$

$$= \int_A \left(\prod_{i=0}^{s-1} \alpha(Z_i)\right) g\left(Z_{\tau_t(B)}\right) \prod_{i=s}^{\tau_t(B)-1} \alpha(Z_i) \mathrm{d}P_z$$

$$\stackrel{A \in \mathcal{A}_s}{=} \int_A \left(\prod_{i=0}^{s-1} \alpha(Z_i)\right) \mathrm{E}\left(g\left(Z_{\tau_t(B)}\right) \prod_{i=s}^{\tau_t(B)-1} \alpha(Z_i) \Big| \mathcal{A}_s\right) \mathrm{d}P_z$$

$$\stackrel{\text{M.C.}}{=} \int_A \left(\prod_{i=0}^{s-1} \alpha(Z_i)\right) \mathrm{E}\left(g\left(Z'_{\tau'_{t-s}(B)}\right) \prod_{i=0}^{\tau'_{t-s}(B)-1} \alpha(Z'_i) \Big| Z'_0 = Z_s\right) \mathrm{d}P_z$$

$$\stackrel{Z_s \in B^{*D}}{\underset{t-s \in D}{\leqslant}} \int_A \left(\prod_{i=0}^{s-1} \alpha(Z_i)\right) g(Z_s) \mathrm{d}P_z$$

$$\stackrel{\text{def. of } X_s}{=} \int_A X_s \mathrm{d}P_z.$$

$\square$

Remark: The assumption "$t - s \in D$" is necessary as shown in Example 7.8 (page 76).

7.8 Example: Necessities for Partial Improving II

This example will show the necessity of the condition "$t - s \in D$" for (7.7.1).

Imagine a Markov chain with seven states a, b, c, d, e, f, h and the following transition matrix

	a	b	c	d	e	f	h
a	0	1	0	0	0	0	0
b	0	0	1	0	0	0	0
c	0	0	0	1	0	0	0
d	0	0	0	0	1	0	0
e	0	0	0	0	0	1	0
f	0	0	0	0	0	0	1
h	0	1	0	0	0	0	0

as drawn in the following picture

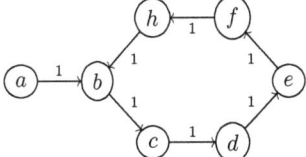

and the payments, i.e. the function g

state	payment
a	4
b	2
c	7
d	5
e	5
f	3
h	8

and we will use $D := \{1, 3\}$ and $\alpha :\equiv 1$. We initialize the algorithm with
$$B := \{a, b, c, d, e, f, h\}.$$
Hence we have
$$B^{*D} = \{c, e, h\}$$
and
$$F = \{h\}.$$
Now consider $\rho :\equiv 0$, $A := \{Z_0 \in \{d\}\}$, $t := 3$, $s := 1$.

We have $A \in \mathcal{A}_1$ and $A \subseteq \{Z_1 = e\} \subseteq \{Z_1 \in B^{*D}\} = \{Z_s \in B^{*D}\}$
(With the definitions of the proof of (7.4.5) and (7.9.4) below we have $A^{0,3} = \{Z_0 \in \{a, d\}\}$
splitting into the sets $A = A_1 = \{Z_0 \in \{d\}\}$ and $A_2 = \{Z_0 \in \{a\}\}$.)

We have $t - s = 3 - 1 = 2 \notin D$ and

$$\begin{aligned}
& \mathrm{E}\left(\mathbf{1}_A X_{\tau_t(B)} \middle| Z_0 = d\right) \\
={} & \mathrm{E}\left(\mathbf{1}_A X_{\tau_3(B)} \middle| Z_0 = d\right) \\
={} & \mathrm{E}\left(X_{\tau_3(B)} \middle| Z_0 = d\right) \\
={} & \mathrm{E}\left(X_3 \middle| Z_0 = d\right) \\
={} & \mathrm{E}\left(g(Z_3) \middle| Z_0 = d\right) \\
={} & g(h) \\
={} & 8 \\
>{} & 5 \\
={} & g(e) \\
={} & \mathrm{E}\left(X_1 \middle| Z_0 = d\right) \\
={} & \mathrm{E}\left(\mathbf{1}_A X_1 \middle| Z_0 = d\right) \\
={} & \mathrm{E}\left(\mathbf{1}_A X_s \middle| Z_0 = d\right),
\end{aligned}$$

contradicting (7.7.1) and (7.9.8).

7.9 Proof of Lemma 7.4 (page 73)

Proof of (7.4.2), (7.4.3), (7.4.4): We have
$$\rho = \hat{\rho} = \tau_\sigma(B^{*D}) \text{ on } \{\rho = \tau_\sigma(B^{*D})\} = \{\rho = \infty\} \cup \{Z_\rho \in B^{*D}\}.$$
Furthermore we have $B^{*\emptyset} = B$,

(7.9.1) $$B^{*D} = \bigcap_{j \in D}^{\kappa} B^{*D \leqslant j}.$$

We have
$$\{Z_n \notin B^{*D}\} \stackrel{(7.9.1)}{=} \bigcup_{j \in D} \{j = \inf\{i \in D \,;\, Z_n \notin B^{*D \leqslant i}\}\} \text{ for all } n \in \mathbb{N}_0.$$
So for all $n \in \mathbb{N}_0$ and $j \in D$ we have
$$\rho < \rho + 1 = n + 1 \leqslant \overbrace{(\tau_{n+j}(B) \wedge \tau_\sigma(B^{*D}))}^{=\hat{\rho}} \leqslant \tau_\sigma(B^{*D})$$
$$\text{on } \{\rho = n\} \cap \{j = \inf\{i \in D \,;\, Z_n \notin B^{*D \leqslant i}\}\}.$$
Hence

(7.9.2) $$\{\rho < \tau_\sigma(B^{*D})\} = \{\rho < \infty\} \cap \{Z_\rho \notin B^{*D}\} = \bigcup_{n=0}^{\infty} \{\rho = n\} \cap \{Z_n \notin B^{*D}\}. \qquad \square$$

Proof of (7.4.5): Let $z \in S$ and define
$$A := \{\rho < \tau_\sigma(B^{*D})\} \stackrel{(7.9.2)}{=} \{\rho < \infty\} \cap \{Z_\rho \notin B^{*D}\}.$$
We have
$$A^c = \{\rho = \infty\} \cup \{Z_\rho \in B^{*D}\} = \{\rho = \tau_\sigma(B^{*D})\} \subseteq \{\rho = \hat{\rho}\}$$
and thus obviously
$$\mathrm{E}\left(\mathbf{1}_{A^c} X_\rho \middle| Z_0 = z\right) \leqslant \mathrm{E}\left(\mathbf{1}_{A^c} X_{\hat{\rho}} \middle| Z_0 = z\right).$$
Define for all $n \in \mathbb{N}_0$ and $j \in D$
$$A^{n,j} := \{\rho = n\} \cap \{j = \inf\{i \in D \,;\, Z_n \notin B^{*D \leqslant i}\}\}.$$
We have
$$A = \biguplus_{n=0}^{\infty} \biguplus_{j \in D} A^{n,j}.$$
For all $n \in \mathbb{N}_0$ and $j \in D$ we have by Lemma 7.6 (page 75)

(7.9.3) $$\mathrm{E}\left(\mathbf{1}_{A^{n,j}} X_\rho \middle| Z_0 = z\right) = \mathrm{E}\left(\mathbf{1}_{A^{n,j}} X_n \middle| Z_0 = z\right) \leqslant \mathrm{E}\left(\mathbf{1}_{A^{n,j}} X_{\tau_{n+j}(B)} \middle| Z_0 = z\right).$$

Due to dominated convergence it suffices to show for all $n \in \mathbb{N}_0$ and $j \in D$ the inequality

(7.9.4) $$\mathrm{E}\left(\mathbf{1}_{A^{n,j}} X_{\tau_{n+j}(B)} \middle| Z_0 = z\right) \leqslant \mathrm{E}\left(\mathbf{1}_{A^{n,j}} X_{\hat{\rho}} \middle| Z_0 = z\right). \qquad \square$$

Remark:

- (7.9.3) stays correct even for arbitrary $D \subseteq \mathbb{N}$.
- As shown in Example 7.8 (page 76) the inequality (7.9.4) may be false if there is no $k \in \mathbb{N} \cup \{\infty\}$ with $D = \mathbb{N}_{\leqslant k}$.
- Nevertheless it seems that the inequality

$$(7.9.5) \qquad \mathrm{E}\left(\mathbf{1}_{A^{n,j}} X_\rho | Z_0 = z\right) \leqslant \mathrm{E}\left(\mathbf{1}_{A^{n,j}} X_{\hat{\rho}} | Z_0 = z\right).$$

is true for all D with $1 \in D$ as studies with our computer program show, but this could not be proved.

Proof of (7.9.4): Define
$$E := \left\{\tau_{n+j}(B) \leqslant \tau_\sigma(B^{*D})\right\}.$$

We have
$$\tau_{n+j}(B) = \tau_{n+j}(B) \wedge \tau_\sigma(B^{*D}) = \hat{\rho} \text{ on } A^{n,j} \cap E,$$

hence

$$(7.9.6) \qquad \mathrm{E}\left(\mathbf{1}_{A^{n,j} \cap E} X_{\tau_{n+j}(B)} \big| Z_0 = z\right) \leqslant \mathrm{E}\left(\mathbf{1}_{A^{n,j} \cap E} X_{\hat{\rho}} \big| Z_0 = z\right).$$

For all $m \in \mathbb{N}$ with $m < j$ define

$$(7.9.7) \qquad A_m := A^{n,j} \cap \{Z_{n+m} \in B^{*D}\} \cap \bigcap_{l=1}^{m-1} \{Z_{n+l} \notin B^{*D}\}.$$

For all $m \in \mathbb{N}$ with $m < j$ and $j - m \in D$ we have by Lemma 7.7 (page 76) (with setting $A := A_m$, $s := n+m$, $t := n+j$)

$$(7.9.8) \qquad \int_{A_m} X_{\tau_{n+j}(B)} \mathrm{d}P_z \overset{7.7}{\leqslant} \int_{A_m} X_{n+m} \mathrm{d}P_z \overset{A_m \subseteq \{n+m = \tau_\sigma(B^{*D})\}}{=} \int_{A_m} X_{\tau_\sigma(B^{*D})} \mathrm{d}P_z \overset{A_m \subseteq E^c}{=} \int_{A_m} X_{\hat{\rho}} \mathrm{d}P_z.$$

We will prove next

$$(7.9.9) \qquad \left(A^{n,j} \cap E^c\right) = \biguplus_{m=1}^{j-1} A_m.$$

Due to linearity of expectation this finishes the proof. □

Proof of (7.9.9):
Disjoint: For all $m, l \in \mathbb{N}$ with $m, l < j$, $m \neq l$ we have

$$\emptyset \subseteq A_m \cap A_l \subseteq \{Z_{n+m} \in B^{*D}\} \cap \{Z_{n+m} \notin B^{*D}\} = \emptyset, \text{ hence } A_m \cap A_l = \emptyset.$$

"⊇": For all $m \in \mathbb{N}$ with $m < j$ we have due to $n + m < n + j \leqslant \tau_{n+j}(B)$

$$A_m \overset{(7.9.7)}{\subseteq} \{\tau_\sigma(B^{*D}) = n+m\} \subseteq \{\tau_\sigma(B^{*D}) < \tau_{n+j}(B)\} \overset{\text{def.}}{=} E^c.$$

"$\subseteq$": Consider $\omega \in A^{n,j} \cap E^c$.
For all $l \in \mathbb{N}_0$ with $n+j \leqslant l < \tau_{n+j}(B)(\omega) = \inf\{l \geqslant n+j \,;\, Z_l(\omega) \in B\}$ we obviously have $Z_l(\omega) \notin B$ and due to $B^{*D} \subseteq B$ we have $Z_l(\omega) \notin B^{*D}$.
Since we consider
$$\omega \in A^{n,j} \cap E^c \subseteq \{\tau_\sigma(B^{*D}) < \tau_{n+j}(B)\} \cap \{\rho = n\},$$
hence
$$\tau_{n+j}(B)(\omega) > \tau_\sigma(B^{*D})(\omega) = \inf\{l \geqslant \sigma(\omega)\,;\, Z_l(\omega) \in B^{*D}\},$$
we obtain $n+j > \tau_\sigma(B^{*D})(\omega)$ and by (7.4.1) $\tau_\sigma(B^{*D})(\omega) \geqslant \rho(\omega) = n$. By the definition of $A^{n,j}$ we have $Z_n(\omega) \notin B^{*D_{\leqslant j}} \supseteq B^{*D}$, hence $\tau_\sigma(B^{*D})(\omega) \neq n$. So we have $n < \tau_\sigma(B^{*D})(\omega) < n+j$, thus it follows $\omega \in A_{[\tau_\sigma(B^{*D})(\omega)-n]}$ by (7.9.7). □

Proof of (7.4.7): Define $\sigma_0 := \sigma$ and inductively
$$\sigma_{k+1} := \hat{\sigma}_k \text{ for all } k \in \mathbb{N}_0.$$

First we will prove by induction that
(7.9.10) $\qquad \sigma_k \leqslant \sigma_{k+1} \leqslant \tau_\sigma(B^{*D})$ for all $k \in \mathbb{N}_0$.

We have $\sigma = \sigma_0 \leqslant \tau_\sigma(B^{*D})$.
Then we have $\sigma = \sigma_0 \leqslant \sigma_1 \leqslant \tau_\sigma(B^{*D})$ by (7.4.2) and (7.4.3).
Consider $k \in \mathbb{N}_0$ with $\sigma \leqslant \sigma_k \leqslant \tau_\sigma(B^{*D})$.
It follows $\sigma \leqslant \sigma_k \leqslant \sigma_{k+1} \leqslant \tau_\sigma(B^{*D})$ by (7.4.2) and (7.4.3).

By (7.9.10) we have
$$\lim_{k\to\infty} \sigma_k \leqslant \tau_\sigma(B^{*D})$$
and by (7.4.4)
$$\sigma_k + 1 \leqslant \sigma_{k+1} \text{ on } \{\sigma_k < \tau_\sigma(B^{*D})\} \text{ for all } k \in \mathbb{N},$$
hence we have
$$\sigma_0 \leqslant \sigma_1 \leqslant \sigma_2 \leqslant \ldots \leqslant \lim_{k\to\infty} \sigma_k = \tau_\sigma(B^{*D}).$$

Let $z \in S$. We have for all $k \in \mathbb{N}_0$
(7.9.11) $\qquad \mathrm{E}\left(X_{\sigma_k}|Z_0 = z\right) \leqslant \mathrm{E}\left(X_{\sigma_{k+1}}|Z_0 = z\right) \qquad$ by (7.4.5).

We infer by induction and assuming (7.4.6)
$$\begin{aligned}
\mathrm{E}\left(X_\sigma|Z_0 = z\right) &= \mathrm{E}\left(X_{\sigma_0}|Z_0 = z\right) \\
&\leqslant \lim_{k\to\infty} \mathrm{E}\left(X_{\sigma_k}|Z_0 = z\right) && \text{by (7.9.11)} \\
&= \mathrm{E}\left(X_{\lim_{k\to\infty}\sigma_k}\Big|Z_0 = z\right) && \text{by (7.4.6)} \\
&= \mathrm{E}\left(X_{\tau_\sigma(B^{*D})}\big|Z_0 = z\right).
\end{aligned}$$
□

7.10 Remark

We used the superscript $*D$ above to depict that some parts of the proofs and lemmatas are true regardless of the structure of D, but we could show the following results only for the special form of D being an initial sequence of $\mathbb{N}$. Therefore we defined above in Setting 7.2 (page 72) for all $k \in \mathbb{N}$ the shortcut B^{*k} for $B^{*\mathbb{N}_{\leqslant k}}$.

7.11 Theorem

Let $\kappa : \mathbb{N} \longrightarrow \mathbb{N} \cup \{\infty\}$ and $B^0 \in \mathcal{S}$.

Define inductively

(7.11.1) $$B^k := (B^{k-1})^{*\kappa(k)} \text{ for all } k \in \mathbb{N}, \quad F := \bigcap_{k \in \mathbb{N}_0} B^k.$$

Assume (as above)

(7.4.6) $$\mathrm{E}\left(X_{\lim_{n \to \infty} \sigma_n} \big| Z_0 = z\right) = \lim_{n \to \infty} \mathrm{E}\left(X_{\sigma_n} \big| Z_0 = z\right)$$
for all $z \in S$ and all non-decreasing sequences $(\sigma_n)_{n \in \mathbb{N}_0}$ of stopping rules.

Then we have

(7.11.2) $\tau_0(B^k) \leqslant \tau_0(B^{k+1})$ for all $k \in \mathbb{N}_0$,

(7.11.3) $\lim_{k \to \infty} \tau_0(B^k) = \tau_0(F)$,

(7.11.4) $\mathrm{E}\left(X_{\tau_0(B^k)} \big| Z_0 = z\right) \leqslant \mathrm{E}\left(X_{\tau_0(B^{k+1})} \big| Z_0 = z\right)$ for all $k \in \mathbb{N}_0$ for all $z \in S$,

(7.11.5) $\lim_{k \to \infty} \mathrm{E}\left(X_{\tau_0(B^k)} \big| Z_0 = z\right) = \mathrm{E}\left(X_{\tau_0(F)} \big| Z_0 = z\right) = \sup_{\tau \in \mathcal{T}(B^0)} \mathrm{E}(X_\tau | Z_0 = z)$ for all $z \in S$,

hence $\tau_0(F)$ is optimal in $\mathcal{T}(B^0)$.

Remark: The herein made assumption is the same as in Lemma 7.4 (page 73).

Remark: For getting the optimal stopping rule, set $B^0 = S$.

Proof of (7.11.2) and (7.11.4): Let $k \in \mathbb{N}_0$. By Lemma 7.4 (page 73) we have
$$B^{k+1} = (B^k)^{*\kappa(k)} \supseteq B^k$$
and
$$\tau_0(B^k) \leqslant \tau_{\tau_0(B^k)}((B^k)^{*\kappa(k)}) = \tau_{\tau_0(B^k)}(B^{k+1}) = \tau_0(B^{k+1}).$$
(Set B and σ of the cited lemma to B^k and $\tau_0(B^k)$.)

Hence we can infer (7.11.2) and (7.11.4). □

Proof of (7.11.3): (7.11.3) follows by the definitions in (7.11.1) and (7.2.2). □

Proof of (7.11.5):
(1): At first we will show that for any $\rho \in \mathcal{T}(B^0)$ there exists
$\rho^\infty \in \mathcal{T}(F)$ with $\rho^\infty \geqslant \rho$ and $\mathrm{E}(X_\rho | Z_0 = z) \leqslant \mathrm{E}(X_{\rho^\infty} | Z_0 = z)$ for all $z \in S$.
Let $\rho \in \mathcal{T}(B^0)$. Define
$$\rho^k := \inf\{n \geqslant \rho ; Z_n \in B^k\} \text{ for all } k \in \mathbb{N}_0,$$
$$\rho^\infty := \inf\{n \geqslant \rho ; Z_n \in F\}.$$

Since $(B^k)_{k\in\mathbb{N}_0}$ is a non-increasing sequence of sets with limit F, $(\rho^k)_{k\in\mathbb{N}_0}$ is a non-decreasing sequence of stopping rules with limit ρ^∞. Let $k \in \mathbb{N}_0$.
By Lemma 7.4 (page 73) we have

$$\mathrm{E}\left(X_{\rho^k}|Z_0 = z\right) \leq \mathrm{E}\left(X_{\rho^{k+1}}|Z_0 = z\right) \text{ for all } z \in S.$$

(Set B and σ of the cited lemma to B^k and ρ^k, then we have $\tau_{\rho^k}((B^k)^{**\kappa(k)}) = \rho^{k+1}$.)

Since $\rho \in \mathcal{T}(B^0)$, we have $\rho = \rho^0$.
So it follows

$$\mathrm{E}(X_\rho|Z_0 = z) \leq \lim_{k\to\infty} \mathrm{E}\left(X_{\rho^k}|Z_0 = z\right)$$
$$\stackrel{(7.4.6)}{\leq} \mathrm{E}\left(X_{\lim_{k\to\infty} \rho^k}|Z_0 = z\right) = \mathrm{E}(X_{\rho^\infty}|Z_0 = z) \text{ for all } z \in S.$$

---

(2): Let $\rho, \tau \in \mathcal{T}(F)$ such that $\rho \leq \tau$. We will show

$$\mathrm{E}(X_\tau|Z_0 = z) \leq \mathrm{E}(X_\rho|Z_0 = z) \text{ for all } z \in S.$$

Define

$$\rho_k := \rho \mathbf{1}_{\{\rho=\tau\}} + \mathbf{1}_{\{\rho<\tau\}} \sum_{n=0}^{\infty} \mathbf{1}_{\{\rho=n\}} \tau_{n+1}(B^k) \text{ for all } k \in \mathbb{N}_0,$$

$$\rho_\infty := \rho \mathbf{1}_{\{\rho=\tau\}} + \mathbf{1}_{\{\rho<\tau\}} \sum_{n=0}^{\infty} \mathbf{1}_{\{\rho=n\}} \tau_{n+1}(F).$$

Then we have $\rho \leq \rho_k \leq \rho_{k+1} \leq \rho_\infty \leq \tau$ for all $k \in \mathbb{N}_0$ and $\lim_{k\to\infty} \rho_k = \rho_\infty$.
Furthermore we have

$$\rho + 1 = n + 1 \leq \tau_{n+1}(F) = \rho_\infty \text{ on } \{\rho < \tau\} \cap \{\rho = n\} \text{ for all } n \in \mathbb{N}_0,$$

hence

(7.11.6) $$\rho + 1 \leq \rho_\infty \text{ on } \biguplus_{n=0}^{\infty} \{\rho = n\} \cup \{\rho < \tau\} = \{\rho \neq \tau\}.$$

Let $(Z'_n)_{n\in\mathbb{N}_0}$ be an independent copy of $(Z_n)_{n\in\mathbb{N}_0}$ and for all $p \in \mathbb{N}_0$ and $B \in \mathcal{S}$ let $\tau'_p(B) = \inf\{q \geq p\,;\, Z'_q \in B\}$.
Let $n \in \mathbb{N}_0$ and $k \in \mathbb{N}_0$.
Since $\rho \in \mathcal{T}(F)$ we have $Z_\rho \in F$ on $\{\rho < \infty\}$ and thus

$$\{\rho = n\} \cap \{\rho < \tau\} \subseteq \{Z_n \in F\}$$
$$\subseteq \{Z_n \in (B^k)^{**\kappa(k)}\}$$
$$\subseteq \left\{g(Z_n) \geq \mathrm{E}\left(\left(\prod_{i=0}^{\tau'_1(B^k)-1} \alpha(Z'_i)\right) g(Z'_{\tau'_{n+1}(B^k)}) \middle| Z'_0 = Z_n\right)\right\},$$

hence we have for all $z \in S$

$$\int_{\{\rho=n<\tau\}} X_{\rho_k} \mathrm{d}P_z = \int_{\{\rho=n<\tau\}} \left(\prod_{i=0}^{\tau_{n+1}(B^k)-1} \alpha(Z_i) \right) g(Z_{\tau_{n+1}(B^k)}) \mathrm{d}P_z$$

$$= \int_{\{\rho=n<\tau\}} \mathrm{E} \left(\left. \left(\prod_{i=0}^{\tau'_1(B^k)-1} \alpha(Z'_i) \right) g(Z'_{\tau'_{n+1}(B^k)}) \right| Z'_0 = Z_n \right) \prod_{i=0}^{n-1} \alpha(Z_i) \mathrm{d}P_z$$

$$\leq \int_{\{\rho=n<\tau\}} g(Z_n) \prod_{i=0}^{n-1} \alpha(Z_i) \mathrm{d}P_z$$

$$= \int_{\{\rho=n<\tau\}} X_\rho \mathrm{d}P_z.$$

By dominated convergence follows

$$\mathrm{E}\left(X_{\rho_k}|Z_0=z\right) \leq \mathrm{E}\left(X_\rho|Z_0=z\right) \text{ for all } z \in S.$$

Letting $k \to \infty$ we obtain

(7.11.7) $\qquad \mathrm{E}\left(X_{\rho_\infty}|Z_0=z\right) \leq \mathrm{E}\left(X_\rho|Z_0=z\right) \text{ for all } z \in S.$

Define $\rho^0 = \rho$ and $\rho^k = (\rho^{k-1})_\infty$ for all $k \in \mathbb{N}$. Then obviously $\rho \leq \rho^k \leq \rho^{k+1} \leq \tau$, and $\rho^k \in \mathcal{T}(F)$ for all $k \in \mathbb{N}$. Hence we have by (7.11.6) $\lim_{k \to \infty} \rho^k = \tau$.
It follows for all $z \in S$

$$\mathrm{E}\left(X_\tau|Z_0=z\right) = \mathrm{E}\left(X_{\lim_{k \to \infty} \rho^k}\middle|Z_0=z\right) \stackrel{(7.4.6)}{=} \lim_{k \to \infty} \mathrm{E}\left(X_{\rho^k}|Z_0=z\right) \stackrel{(7.11.7)}{\leq} \mathrm{E}\left(X_\rho|Z_0=z\right).$$

---

(3): Let $\sigma \in \mathcal{T}(B^0)$. By (1), there exists $\tau \in \mathcal{T}(F)$ with $\tau \geq \sigma$ such that

$$\mathrm{E}\left(X_\sigma|Z_0=z\right) \leq \mathrm{E}\left(X_\tau|Z_0=z\right) \text{ for all } z \in S.$$

Obviously we have $\tau_0(F) \leq \tau$ due to $\tau \in \mathcal{T}(F)$. Hence it follows by (2)

$$\mathrm{E}\left(X_\sigma|Z_0=z\right) \leq \mathrm{E}\left(X_\tau|Z_0=z\right) \leq \mathrm{E}\left(X_{\tau_0(F)}\middle|Z_0=z\right) \text{ for all } z \in S. \qquad \square$$

8 Markovian Case with Random Discounting and Finite State Space

In this chapter we consider the Markovian case with random discounting as discussed in Section 7 (page 71) and assume additionally finite state space. The algorithm requires the calculation of many conditional expectations. We show in this section that these expectations can be calculated by solving linear equations under the given assumptions. Numerical examples can be found in Section 9 (page 89).

8.1 Setting

We continue with the setting of Setting 7.2 (page 72) and additionally assume finite S. Let Π be the transition matrix and define Ψ by $\Psi_{z,y} := \alpha(z) \cdot \Pi_{z,y}$ for all $z, y \in S$. Define for all $p \in \mathbb{N}$ and all $B \in \mathcal{S}$

$$h'_{B,p} : S \longrightarrow \mathbb{R}, \; z \longmapsto \mathrm{E}\left(X_{\tau_p(B)} \big| Z_0 = z\right) = \mathrm{E}\left(g\left(Z_{\tau_p(B)}\right) \prod_{j=0}^{\tau_p(B)-1} \alpha(Z_i) \bigg| Z_0 = z\right).$$

8.2 Theorem

For all $B \in \mathcal{S}$, $p \in \mathbb{N}$ and $z \in S$ we have

$$h'_{B,p}(z) = \sum_{y \in S} (\Psi^p)_{z,y} \, h'_{B,0}(y).$$

Proof: Let $B \in \mathcal{S}$.
We have for all $p \in \mathbb{N}$ and $z \in S$

$$
\begin{aligned}
h'_{B,p}(z) &= \mathrm{E}\left(g\left(Z_{\tau_p(B)}\right)\prod_{j=0}^{\tau_p(B)-1}\alpha(Z_i)\bigg|Z_0=z\right) \\
&= \alpha(z)\mathrm{E}\left(g\left(Z_{\tau_p(B)}\right)\prod_{j=1}^{\tau_p(B)-1}\alpha(Z_i)\bigg|Z_0=z\right) \\
&= \alpha(z)\sum_{y\in S}P\left(Z_1=y\mid Z_0=z\right)\mathrm{E}\left(g\left(Z_{\tau_p(B)}\right)\prod_{j=1}^{\tau_p(B)-1}\alpha(Z_i)\bigg|Z_1=y, Z_0=z\right) \\
&\stackrel{\mathrm{M.C.}}{=} \sum_{y\in S}\alpha(z)P\left(Z_1=y\mid Z_0=z\right)\mathrm{E}\left(g\left(Z_{\tau_{p-1}(B)}\right)\prod_{j=0}^{\tau_{p-1}(B)-1}\alpha(Z_i)\bigg|Z_0=y\right) \\
&= \sum_{y\in S}\alpha(z)\Pi_{z,y}\mathrm{E}\left(g\left(Z_{\tau_{p-1}(B)}\right)\prod_{j=0}^{\tau_{p-1}(B)-1}\alpha(Z_i)\bigg|Z_0=y\right) \\
&= \sum_{y\in S}\Psi_{z,y}\mathrm{E}\left(g\left(Z_{\tau_{p-1}(B)}\right)\prod_{j=0}^{\tau_{p-1}(B)-1}\alpha(Z_i)\bigg|Z_0=y\right) \\
&= \sum_{y\in S}\Psi_{z,y}h'_{B,p-1}(y).
\end{aligned}
$$

Hence we have
$$h'_{B,p} = \Psi \cdot h'_{B,p-1}.$$
By induction we have for all $p \in \mathbb{N}$
$$h'_{B,p} = \Psi^p \cdot h'_{B,0},$$
hence for all $p \in \mathbb{N}$ and $z \in S$ we have
$$h'_{B,p}(z) = \sum_{y\in S}(\Psi^p)_{z,y}h'_{B,0}(y).$$
□

8.3 Corollar

Using the definitions of Setting 7.2 (page 72) we have for all $B \in \mathcal{S}$ and $D \subseteq \mathbb{N}$

$$B^{*D} = \left\{z \in B\ ;\ g(z) \geq \sup_{p\in D}\sum_{y\in S}(\Psi^p)_{z,y}h'_{B,0}(y)\right\}.$$

Proof: Due to the definitions in Setting 8.1 (page 85) the iterative step is performed for all $B \in \mathcal{S}$ and $D \subseteq \mathbb{N}$ by

$$B^{*D} = \left\{z \in B\ ;\ g(z) \geq \sup_{p\in D}h'_{B,p}(z)\right\},$$

so the statement follows by Theorem 8.2 (page 85). □

8.4 FII by Solving Linear Equations[1]

Let $B \in \mathcal{S}$ and assume

(8.4.1) $\qquad$ for all $z \in S$ $([\alpha(z) < 1]$ or $[P(\tau_0(B) < \infty \mid Z_0 = z) = 1])$.

Then $h'_{B,0}$ is the unique solution of the system of linear equations which is constructed in the following:

Define the vector b by
$$b_i := \begin{cases} 1 & i \in B, \\ 0 & i \notin B, \end{cases} \quad \text{for all } i \in S,$$

the vector d by
$$d_i := b_i \cdot g(i) \text{ for all } i \in S$$

and the matrix A by
$$A_{i,j} := \delta_{i,j} - (1 - b_i)\Psi_{i,j} \text{ for all } i, j \in S.$$

Proof: For all $h \in \mathbb{R}^S$ we have by definition
$$d = Ah \iff \forall z \in B \ h(z) = g(z) \land \forall z \notin B \ h(z) = \alpha(z) \sum_{y \in S} P(Z_1 = y \mid Z_0 = z) h(y).$$

Now we prove in two steps that $\{h \in \mathbb{R}^S \ ; \ d = Ah\} = \{h'_{B,0}\}$.

Proof of $\{h'_{B,0}\} \subseteq \{h \in \mathbb{R}^S \ ; \ d = Ah\}$:

For all $z \in B$ we have $Z_0 = z \in B$ and thus $\tau_0(B) = 0$ and $\prod_{j=0}^{\tau_0(B)-1} \alpha(Z_i) = 1$, hence
$$h'_{B,0}(z) = E\left(g(Z_{\tau_0(B)}) \prod_{j=0}^{\tau_0(B)-1} \alpha(Z_i) \Big| Z_0 = z\right) = E(g(Z_0) \cdot 1 | Z_0 = z) = g(z)$$

and for all $z \notin B$ we have
$$\begin{aligned}
h'_{B,0}(z) &= E\left(g(Z_{\tau_0(B)}) \prod_{j=0}^{\tau_0(B)-1} \alpha(Z_i) \Big| Z_0 = z\right) \\
&= \alpha(z) E\left(g(Z_{\tau_0(B)}) \prod_{j=1}^{\tau_0(B)-1} \alpha(Z_i) \Big| Z_0 = z\right) \\
&= \alpha(z) \sum_{y \in S} P(Z_1 = y \mid Z_0 = z) E\left(g(Z_{\tau_1(B)}) \prod_{j=1}^{\tau_1(B)-1} \alpha(Z_i) \Big| Z_1 = y\right) \\
&\stackrel{\text{M.C.}}{=} \alpha(z) \sum_{y \in S} P(Z_1 = y \mid Z_0 = z) E\left(g(Z_{\tau_0(B)}) \prod_{j=0}^{\tau_0(B)-1} \alpha(Z_i) \Big| Z_0 = y\right) \\
&= \alpha(z) \sum_{y \in S} P(Z_1 = y \mid Z_0 = z) h'(y).
\end{aligned}$$

---
[1][Irl09, part 2 and part 3]

So we have $d = Ah'_{B,0}$.

Proof of $\left|\{h \in \mathbb{R}^S \ ; \ d = Ah\}\right| = 1$:
By (8.4.1) we have

$$\left(\left(\alpha(z) < 1\right) \text{ or } \left(z \notin B \ \Rightarrow \ \sum_{y \notin B} \Pi_{z,y} < 1\right)\right) \text{ for all } z \in S,$$

hence

(8.4.2) $$\sum_{y \notin B} \Psi_{z,y} = \alpha(z) \sum_{y \notin B} \Pi_{z,y} < 1 \text{ for all } z \notin B.$$

We look at the corresponding homogeneous system of linear equations. Hence consider some $h \in \mathbb{R}^S$ with $0 = Ah$. It is obvious that for all $y \in B$ we have $h(y) = 0$. For all $z \notin B$ we have

$$\begin{aligned} h(z) &= \alpha(z) \sum_{y \in S} P(Z_1 = y \mid Z_0 = z) h(y) \\ &= \sum_{y \notin B} \alpha(z) P(Z_1 = y \mid Z_0 = z) h(y) \\ &= \sum_{y \notin B} \Psi_{z,y} h(y). \end{aligned}$$

We have for all $z \notin B$

$$\begin{aligned} |h(z)| = \left|\sum_{y \notin B} \Psi_{z,y} h(y)\right| &\leq \sum_{y \notin B} \Psi_{z,y} |h(y)| \\ &\leq \sum_{y \notin B} \Psi_{z,y} \max\{|h(x)| \ ; \ x \notin B\} \\ &\leq \max\{|h(x)| \ ; \ x \notin B\} \sum_{y \notin B} \Psi_{z,y}. \end{aligned}$$

Hence
$$\max\{|h(x)| \ ; \ x \notin B\} \leq \max\{|h(x)| \ ; \ x \notin B\} \sum_{y \notin B} \Psi_{z,y}.$$

Due to (8.4.2) we have
$$0 = \max\{|h(x)| \ ; \ x \notin B\} \sum_{y \notin B} \Psi_{z,y}.$$

Hence $h \equiv 0$ and the solution for the homogenous system is unique and thus the solution for the inhomogeneous system is unique as well. $\square$

9 Numerical Examples

In the section we give some numerical examples for the Markovian case with constant discounting (see Part 6.3.3 (page 68)) and finite state space using the technique shown in Section 8 (page 85). At first this is done for $\kappa \equiv \{1\}$ in Part 9.2 (page 90) and then (using the definitions of Setting 7.2 (page 72)) for more arbitrary $\kappa : \mathbb{N} \longrightarrow \mathcal{P}ot(\mathbb{N})$ in Part 9.3 (page 92).

9.1 Setting

The examples we focus on in this section are the same as in [Irl06, part 4], but here the technique used is different.

Consider $\{0, ..., 20\} \times \{0, ..., 20\}$ as the state space for the chain and a payoff function g with $g(z, n) = \alpha^n f(z)$ with $f((5,5)) = 10$, $f((5,15)) = f((15,15)) = 0$, $f((x,y)) = 5$ otherwise. The transition probabilities are given in the form

$$p(x+1, y \mid x, y) = 0.5 \cdot p_x, \quad p(x-1, y \mid x, y) = 0.5 \cdot (1 - p_x),$$
$$p(x, y+1 \mid x, y) = 0.5 \cdot p_y, \quad p(x, y-1 \mid x, y) = 0.5 \cdot (1 - p_y),$$

with reflecting boundaries.
For better understanding the following graphic shows the undiscounted pay-offs:

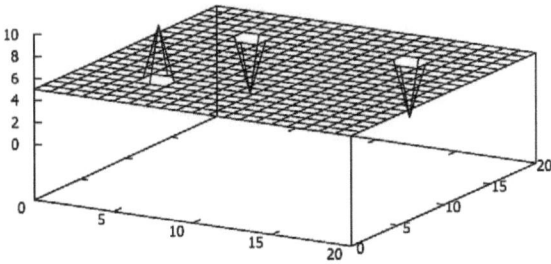

Undiscounted Pay-Offs

9.2 Results for $\kappa \equiv \{1\}$

With a discount-factor $\alpha = 0.98^{1/20}$ our program recommends a stopping rule visible in the next picture. The stopping points are big squares.

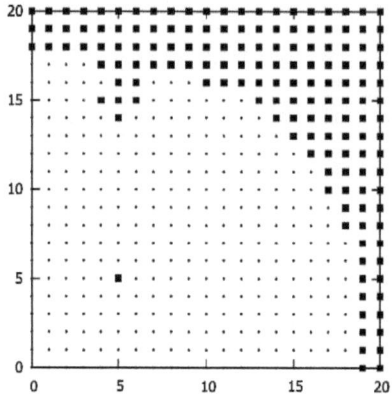

Stopping-Points of the best calculated stopping rule

Using this stopping rule our expected values are the optimal values shown in the next picture:

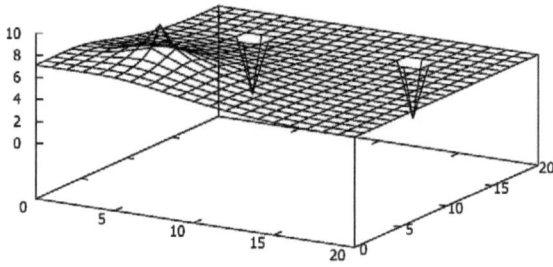

Expected Values

The number of iterations and the time for the calculation depends on κ, but this did not influence the result.
Nevertheless in other examples we found out that due to machine precision the powers of the transition matrix could not be calculated exactly enough, so even the results changed to too small or too big stopping areas.
For constant κ equivalent to $\{1\}$ the program is doing the problem above within 19 iterations and 0.16 seconds.

With $\alpha = 0.99$ and $\kappa \equiv \{1\}$ the following result appears within 6 iterations and 0.04 seconds.

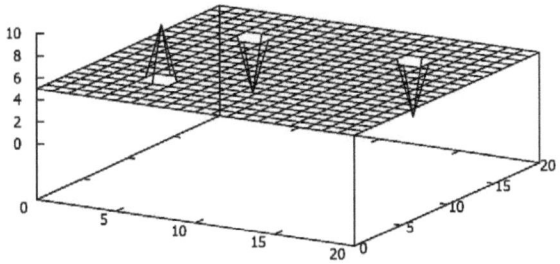

Pay-Offs

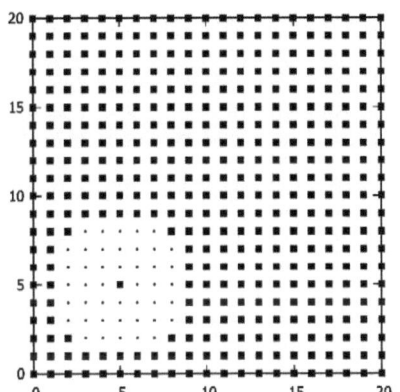

Stopping Points of the Recommended Stopping Rule

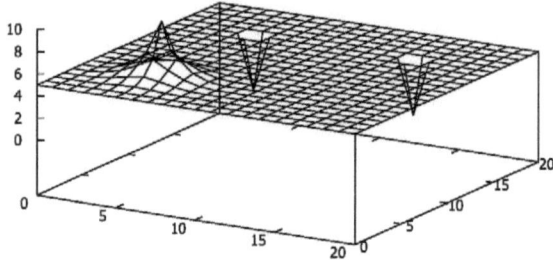

Expected Values, when using the stopping rule above

The results above are similar to those of Irle shown in [Irl06, part 4], where they were calculated by simulation studies.

The speed of the example above is so high that the effect of changing κ cannot be examined therein. Thus we will increase the resolution by factor 10 in each dimension to study the effect of κ, hence we study a state space of $\{0, ..., 200\} \times \{0, ..., 200\}$ with pay-off 10 at $(50, 50)$, pay-off 0 at $(50, 150)$ and $(150, 150)$. We set $\alpha = 0.9999$ so that we have a trivial optimal stopping rule of stopping at $(50, 150)$, $(150, 150)$ and their four sourrounding points as well as at $(50, 50)$ and continuing at all other points.

9.3 Influence of κ

Using the definitions of Setting 7.2 (page 72) we will examine more arbitrary $\kappa : \mathbb{N} \longrightarrow \mathcal{P}ot(\mathbb{N})$.

At first we will examine $\kappa \equiv \mathbb{N}_{\leqslant k}$ with some $k \in \mathbb{N}$ in the text and graphics called "kappa".

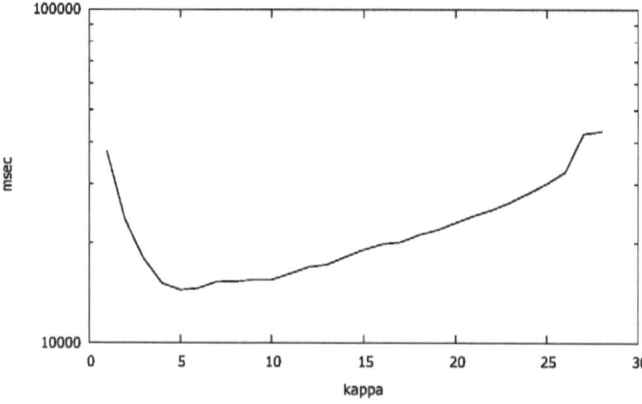

Runtime of the algorithm (in log-scale) plotted against kappa

The largest value of kappa we could examine was 28. At larger values the amount of space needed for storing the powers of the transition matrix explodes and exceeds the RAM-size of 1 GB. The algorithm slows significantly down due to hard disc accesses for reading the matrix powers.

We have a look at the time needed to calculate the powers of the transition matrix. There is a logarithmic dependency on kappa.

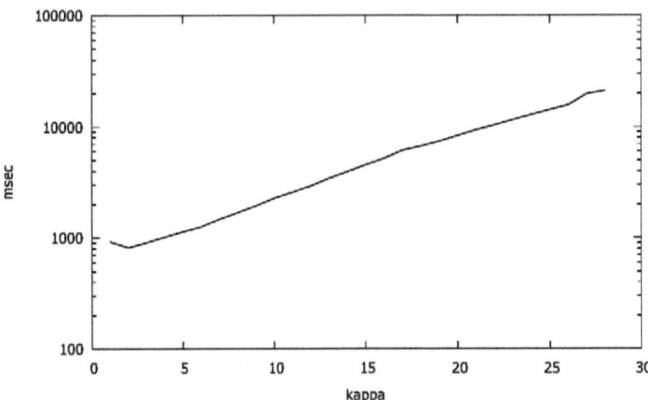

Runtime (in log-scale) for calculating powers of the transition matrix plotted against kappa.

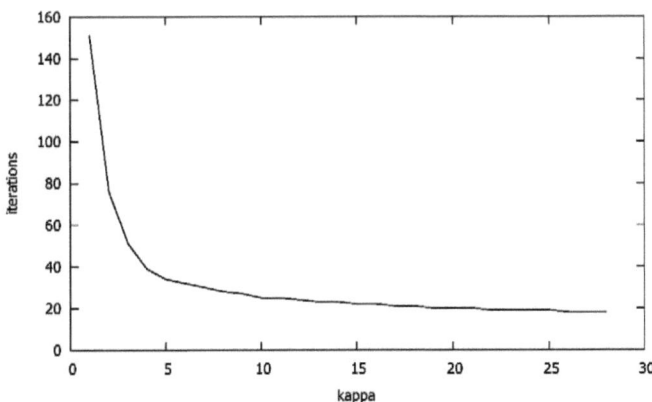

Number of iterations plotted against kappa.

The number of iterations increases with kappa decreasing.

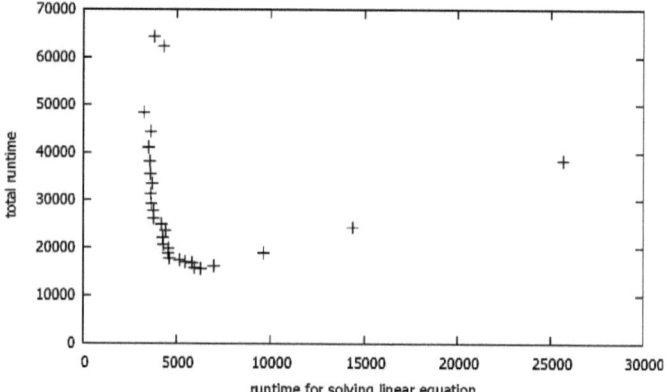

In each iteration we have to solve a linear equation independent of κ and hence independent of kappa. We can see in the example above a more or less linear dependency between the runtime for solving linear equation and the total runtime for the first five points when encountered from right to left. These points are related to kappa going from 1 to 5. The points on the left, more or less one above the other, are related to higher inefficient values of kappa. The number of iterations is strongly correlated with the runtime for solving linear equations and merely related to the exact problem.

We have also done some tests with more arbitrary $\kappa : \mathbb{N} \longrightarrow \mathcal{P}ot(\mathbb{N})$. The time needed for doing the nth iteration does mostly depend on the size of $\kappa(n)$, the set itself does not really influence the runtime.

Consider the algorithm is doing n iterations. If $1 \notin \kappa(n)$ then in our tests the resulting stopping area was often larger then the optimal stopping area, since going one more step was better, but going more than one step was not, but this improvement could not be found by the algorithm, since $1 \notin \kappa(n)$. But given $1 \in \kappa(n)$ (and sufficiently high machine precision) in our tests the final result was always the optimal stopping area.

10 Monotone Case

This chapter examines the term "monotone case" which appears manifold in the literature. By FII Algorithm we can reprove the well known result that in the monotone case the myopic rules are optimal (under weak conditions) and we will enlarge this result to some more stopping rules appearing naturally within FII Algorithm, and adapt the result to the Markovian case.

In Definition 10.1 (page 95) we give an overview over common definitions in this field. We show in Remark 10.2 (page 96) that these terms naturally appear in our model. We enlarge the common terms by the concept of *m-monotone* and *m-stages look-ahead rules* in Definition 10.3 (page 96). Is an m-monotone problem also p-monotone? If $p \geqslant m$ this is true as shown in Lemma 10.4 (page 97), otherwise it is in general not true as shown in Lemma 10.6 (page 98). We show that m-monotone implies optimality of the m-stages look-ahead rules under weak conditions in Theorem 10.8 (page 99). In Part 10.10 (page 100) we give an adaption of this to the Markovian case. (In this chapter we concentrate on the infinite case, but the results can be easily transferred to the finite case.)

10.1 Definition

In [CRS71] and in [Irl79] a problem is called *monotone* iff

$$\{X_n \geqslant \mathrm{E}\,(X_{n+1}|\mathcal{A}_n)\} \subseteq \{X_{n+1} \geqslant \mathrm{E}\,(X_{n+2}|\mathcal{A}_{n+1})\} \text{ for all } n \in \mathbb{N}_0$$

and

$$\bigcup_{n \in \mathbb{N}_0} \{X_n \geqslant \mathrm{E}\,(X_{n+1}|\mathcal{A}_n)\} = \Omega.$$

Further it is called *weakly monotone* in [Irl79] iff

$$\left\{X_n \geqslant \sup_{k \geqslant 1} \mathrm{E}\,(X_{n+k}|\mathcal{A}_n)\right\} \subseteq \left\{X_{n+1} \geqslant \sup_{k \geqslant 1} \mathrm{E}\,(X_{n+1+k}|\mathcal{A}_{n+1})\right\} \text{ for all } n \in \mathbb{N}_0$$

and

$$\bigcup_{n \in \mathbb{N}_0} \left\{X_n \geqslant \sup_{k \geqslant 1} \mathrm{E}\,(X_{n+k}|\mathcal{A}_n)\right\} = \Omega.$$

The stopping rules

$$\min\{n \geqslant l\,;\ X_n \geqslant \mathrm{E}\,(X_{n+1}|\mathcal{A}_n)\}, l \in \mathbb{N}_0$$

are called *one-stage look-ahead rules* or shortly *myopic rules*.

The stopping rules

$$\min\left\{n \geqslant l\,;\ X_n \geqslant \sup_{k \geqslant 1} \mathrm{E}\,(X_{n+k}|\mathcal{A}_n)\right\}, l \in \mathbb{N}_0$$

are called *one-time look-ahead rules*.

10.2 Remark: Appearance in the above Model

The definitions recapitulated in Definition 10.1 (page 95) can be found in the above model. Consider $C^{(0)} \equiv \Omega$ and $\tau^{(0)}$ and $C^{(0)}$ corresponding.

- $\tau^{(1)}$ depends on $\kappa(0)$. If $\kappa(0) = 1$, then it is a family of one-stage look-ahead rules, if $\kappa(0) = \infty$, then it is a family of one-time look-ahead rules.
- $C^{(1)}$ depends also on $\kappa(0)$.
  Given $\kappa(0) = 1$ the problem is monotone iff
  $$\bigcup_{n \in \mathbb{N}_0} C_n^{(1)} = \Omega, \quad C_n^{(1)} \subseteq C_{n+1}^{(1)} \text{ for all } n \in \mathbb{N}_0.$$
  Given $\kappa(0) = \infty$ the problem is weakly monotone iff
  $$\bigcup_{n \in \mathbb{N}_0} C_n^{(1)} = \Omega, \quad C_n^{(1)} \subseteq C_{n+1}^{(1)} \text{ for all } n \in \mathbb{N}_0.$$

10.3 Definition and Remark:
m-monotone and m-stages look-ahead rules

The above remark gives reason to the following definition:
For all $m \in \mathbb{N} \cup \{\infty\}$ we will call a problem m-monotone iff

$$\left\{ X_n \geqslant \sup_{m+1 > k \geqslant 1} \mathrm{E}\left(X_{n+k} | \mathcal{A}_n\right) \right\} \subseteq \left\{ X_{n+1} \geqslant \sup_{m+1 > k \geqslant 1} \mathrm{E}\left(X_{n+1+k} | \mathcal{A}_{n+1}\right) \right\} \text{ for all } n \in \mathbb{N}_0$$

and

$$\bigcup_{n \in \mathbb{N}_0} \left\{ X_n \geqslant \sup_{m+1 > k \geqslant 1} \mathrm{E}\left(X_{n+k} | \mathcal{A}_n\right) \right\} = \Omega.$$

With this definition the expressions "monotone" and "1-monotone" are equivalent as well as the definitions "weakly monotone" and "∞-monotone".

For all $m \in \mathbb{N} \cup \{\infty\}$ we will call the stopping rules

$$\min \left\{ n \geqslant l \; ; \; X_n \geqslant \sup_{m+1 > k \geqslant 1} \mathrm{E}\left(X_{n+k} | \mathcal{A}_n\right) \right\}, l \in \mathbb{N}_0$$

m-stages look-ahead rules.

With this definition the expression "one-time look-ahead rules" becomes equivalent to "infinite-stages look-ahead rules".

Consider $C^{(0)} \equiv \Omega$ and $\tau^{(0)}$ and $C^{(0)}$ corresponding.
Then the problem is $\kappa(0)$-monotone iff

$$\bigcup_{n \in \mathbb{N}_0} C_n^{(1)} = \Omega \quad \text{and} \quad C_n^{(1)} \subseteq C_{n+1}^{(1)} \text{ for all } n \in \mathbb{N}_0,$$

furthermore $\tau^{(1)}$ is a family of $\kappa(0)$-stages look-ahead rules.

10.4 Lemma: m-monotone

Consider $m \in \mathbb{N}$.
If a problem is m-monotone, then for all $p \in \mathbb{N}_0 \cup \{\infty\}$ with $m \leqslant p$ we have

$$\left\{ X_n \geqslant \sup_{m+1 > k \geqslant 1} \mathrm{E}\left(X_{n+k} | \mathcal{A}_n\right) \right\} = \left\{ X_n \geqslant \sup_{p+1 > k \geqslant 1} \mathrm{E}\left(X_{n+k} | \mathcal{A}_n\right) \right\} \text{ for all } n \in \mathbb{N}_0$$

and hence the problem is p-monotone.

Proof: Consider an m-monotone problem, hence we have for all $n \in \mathbb{N}_0$

$$\left\{ X_n \geqslant \sup_{m+1 > k \geqslant 1} \mathrm{E}\left(X_{n+k} | \mathcal{A}_n\right) \right\} \subseteq \left\{ X_{n+1} \geqslant \sup_{m+1 > k \geqslant 1} \mathrm{E}\left(X_{n+1+k} | \mathcal{A}_{n+1}\right) \right\}.$$

Consider $p \in \mathbb{N}_0 \cup \{\infty\}$ with $m \leqslant p$ and $n \in \mathbb{N}_0$.
The inclusion "$\supseteq$" is obvious.
Since the problem is m-monotone we have

$$X_{n+l} \geqslant \mathrm{E}\left(X_{n+l+1} | \mathcal{A}_n\right) \text{ on } \left\{ X_n \geqslant \sup_{m+1 > k \geqslant 1} \mathrm{E}\left(X_{n+k} | \mathcal{A}_n\right) \right\} \text{ for all } l \in \mathbb{N}_0.$$

It follows by induction

$$X_n \geqslant \mathrm{E}\left(X_{n+l} | \mathcal{A}_n\right) \text{ on } \left\{ X_n \geqslant \sup_{m+1 > k \geqslant 1} \mathrm{E}\left(X_{n+k} | \mathcal{A}_n\right) \right\} \text{ for all } l \in \mathbb{N}_0.$$

Thus we have the inclusion "$\subseteq$". □

Hence a monotone problem (1-monotone) is weakly monotone (∞-monotone).

10.5 Conclusion

Consider $C^{(0)} \equiv \Omega$ and $\tau^{(0)}$ and $C^{(0)}$ corresponding.

Consider $m \in \mathbb{N}$ and a m-monotone problem. Then we have for every $\kappa(0) \geqslant m$

$$\bigcup_{n \in \mathbb{N}_0} C_n^{(1)} = \Omega, \quad C_n^{(1)} \subseteq C_{n+1}^{(1)} \text{ for all } n \in \mathbb{N}_0.$$

Proof: For every $\kappa(0) \geqslant m$ we know by Lemma 10.4 (page 97) that the problem is $\kappa(0)$-monotone. The result follows by Definition 10.3 (page 96). □

10.6 Counter-Example: m-monotone

Consider $m, p \in \mathbb{N} \cup \{\infty\}$ with $p < m$.
Being m-monotone does in general not imply being p-monotone.

Proof: Set $l := m$ if $m < \infty$ and $l := p + 1$ otherwise.
Consider e.g. X with $X_1 = 0$, $X_{1+k} = -1$ for all $k \in \mathbb{N}$ with $k < l$, $X_{l+k} = \frac{1}{k}$ for all $k \in \mathbb{N}$.
Hence X is pointwise constant and of the form $(0, -1, -1, ..., -1, -1, 1, \frac{1}{2}, \frac{1}{3}, \frac{1}{4}, ...)$.

For all $n \in \mathbb{N}$ with $n \leqslant l$ we have $n + 1 \leqslant l + 1 \leqslant m + 1 < n + m + 1$ and thus

$$\{0, -1\} \ni X_n < 1 = X_{l+1} \leqslant \sup_{m+1 > k \geqslant 1} X_{n+k} = \sup_{m+1 > k \geqslant 1} \mathrm{E}\left(X_{n+k} | \mathcal{A}_n\right),$$

hence

$$\left\{ X_n \geqslant \sup_{m+1 > k \geqslant 1} \mathrm{E}\left(X_{n+k} | \mathcal{A}_n\right) \right\} = \emptyset.$$

For all $n \in \mathbb{N}$ and $k \in \mathbb{N}$ with $k < m + 1$

$$X_{l+n} = \frac{1}{n} \geqslant \frac{1}{n+k} = X_{l+n+k},$$

thus

$$X_{l+n} \geqslant \sup_{m+1 > k \geqslant 1} X_{l+n+k} = \sup_{m+1 > k \geqslant 1} \mathrm{E}\left(X_{l+n+k} | \mathcal{A}_{l+n}\right),$$

hence

$$\left\{ X_{l+n} \geqslant \sup_{m+1 > k \geqslant 1} \mathrm{E}\left(X_{l+n+k} | \mathcal{A}_{l+n}\right) \right\} = \Omega.$$

Obviously the problem is m-monotone.

We have for all $k \in \mathbb{N}$ with $k < m + 1$

$$\{0, -1\} \ni X_{l-p} \geqslant -1 = X_{l-p+k},$$

hence

$$\sup_{p+1 > k \geqslant 1} X_{l-p+k} = \sup_{p+1 > k \geqslant 1} \mathrm{E}\left(X_{l-p+k} | \mathcal{A}_{l-p}\right)$$

and

$$X_{l-p+1} = -1 < 1 = X_{l+1} = X_{l-p+1+p} = \sup_{p+1 > k \geqslant 1} X_{l-p+1+k} = \sup_{p+1 > k \geqslant 1} \mathrm{E}\left(X_{l-p+1+k} | \mathcal{A}_{l-p+1}\right),$$

hence

$$\Omega = \left\{ X_{l-p} \geqslant \sup_{p+1 > k \geqslant 1} \mathrm{E}\left(X_{l-p+k} | \mathcal{A}_{l-p}\right) \right\}$$

and

$$\emptyset = \left\{ X_{l-p+1} \geqslant \sup_{p+1 > k \geqslant 1} \mathrm{E}\left(X_{l-p+1+k} | \mathcal{A}_{l-p+1}\right) \right\}.$$

Since $\Omega \nsubseteq \emptyset$ the problem is not p-monotone. □

10.7 Lemma

Choose any $\kappa(0)$ and $\kappa(1)$. ($C^{(1)}$ depends on $\kappa(0)$ and $C^{(2)}$ on $\kappa(0)$ and $\kappa(1)$.)
If
$$C_n^{(1)} \subseteq C_{n+1}^{(1)} \text{ for all } n \in \mathbb{N}_0,$$
then we have
$$C^{(1)} = C^{(2)}.$$

Proof: Assume $C_n^{(1)} \subseteq C_{n+1}^{(1)}$ for all $n \in \mathbb{N}_0$. Obviously we have $C_n^{(1)} \subseteq C_{n+k}^{(1)}$ for all $n, k \in \mathbb{N}_0$. Consider $n \in \mathbb{N}_0$. Since $\tau^{(1)}$ and $C^{(1)}$ are corresponding we have by Theorem 2.3 (page 16)
$$C_n^{(1)} \subseteq C_{n+k}^{(1)} = \left\{ \tau_{n+k}^{(1)} = n+k \right\} \text{ for all } k \in \mathbb{N}.$$
Hence on $C_n^{(1)}$ we have
$$\mathrm{E}\left(X_{\tau_{n+k}^{(1)}} \middle| \mathcal{A}_n \right) = \mathrm{E}\left(X_{n+k} \middle| \mathcal{A}_n \right) \text{ for all } k \in \mathbb{N}.$$
We have thus
$$\begin{aligned}
C_n^{(2)} &= C_n^{(1)} \cap \bigcap \left\{ \left\{ \mathrm{E}\left(X_{\tau_p^{(1)}} \middle| \mathcal{A}_n \right) \leqslant X_n \right\} \; ; \; n \leqslant p < \min\{n + \kappa(1), N\} + 1 \right\} \\
&= C_n^{(1)} \cap \bigcap \left\{ \left\{ \mathrm{E}\left(X_p \middle| \mathcal{A}_n\right) \leqslant X_n \right\} \; ; \; n \leqslant p < \min\{n + \kappa(1), N\} + 1 \right\} \\
&= C_n^{(1)} \cap \Omega \cap \bigcap \left\{ C_p^{(1)} \; ; \; n + 1 \leqslant p < \min\{n + \kappa(1), N\} + 1 \right\} \\
&= C_n^{(1)}.
\end{aligned}$$
□

10.8 Theorem

Consider $m \in \mathbb{N} \cup \{\infty\}$, a m-monotone problem and Assumption 5.2 (page 52). Then for each $p \in \mathbb{N} \cup \{\infty\}$ with $p \geqslant m$ the family of p-stages look-ahead rules is optimal.

Proof: Consider $C^{(0)} \equiv \Omega$ and $\tau^{(0)}$ and $C^{(0)}$ corresponding and $\kappa(0) := \kappa(1) := p$. By Conclusion 10.5 (page 97) we have
$$\bigcup_{n \in \mathbb{N}_0} C_n^{(1)} = \Omega, \quad C_n^{(1)} \subseteq C_{n+1}^{(1)} \text{ for all } n \in \mathbb{N}_0.$$
Then we have by Lemma 10.7 (page 99)
$$C^{(1)} = C^{(2)}.$$
By Remark 5.15 (page 64) we have the optimality of $\tau^{(\infty)}$.
Further we have by Conclusion 5.13 (page 63)
$$\tau^{(1)} = \tau^{(\infty)}.$$
□

10.9 Lemma

$\bigcup_{n \in \mathbb{N}_0} C_n^{(1)} = \Omega$ implies $\tau_n^{(1)}$ finite for all $n \in \mathbb{N}_0$.

Proof: Assume there is $n \in \mathbb{N}_0$ and $\omega \in \Omega$ with $\tau_n^{(1)}(\omega) = \infty$. We have

$$\infty = \tau_n^{(1)}(\omega) = \inf\left\{p \in \mathbb{N}_0 \; ; \; \mathbf{1}_{C_p^{(1)}}(\omega) = 1\right\} = \inf\left\{p \in \mathbb{N}_0 \; ; \; \omega \in C_p^{(1)}\right\},$$

hence

$$\emptyset = \left\{p \in \mathbb{N}_0 \; ; \; \omega \in C_p^{(1)}\right\}.$$

So we have $\omega \notin \bigcup_{n \in \mathbb{N}_0} C_n^{(1)} = \Omega$, a contradiction. $\square$

10.10 Markov Case

In Section 6 (page 65) we discussed the Markovian case. As in that section let $(Z_n)_{n \in \mathbb{N}_0}$ be a time-homogeneous Markov process with respect to the underlying filtration with state space $(S, \mathcal{S})$ and the FII algorithm only has to work in the systems of subsets of the state space S.

Assume further that we have a payoff $(X_n)_{n \in \mathbb{N}_0}$ such that our improvement step can be done in the way of Setting 7.2 (page 72) for all $k \in \mathbb{N}$ by

$$S^{*k} \stackrel{\text{def.}}{=} \left\{z \in S \; ; \; g(z) \geqslant \sup\left\{\mathrm{E}\left(X_l | Z_0 = z\right) \; ; \; l \in \mathbb{N}_{\leqslant k}\right\}\right\}.$$

This is the case in all examples within Section 6 (page 65).

For all $k \in \mathbb{N} \cup \{\infty\}$ we will call a problem k-monotone iff

$$\mathrm{P}\left(Z_1 \in S^{*k} | Z_0 = z\right) = 1 \text{ for all } z \in S^{*k},$$

and the k-stages look-ahead rule is $\tau_0(S^{*k})$.

Consider $k \in \mathbb{N}$. The results shown above for the general case in Lemma 10.4 (page 97) and Lemma 10.6 (page 98) can be easily transferred to the Markovian case: A problem being k-monotone is l-monotone if $l \geqslant k$ and in general not l-monotone if $l < k$. If a problem is k-monotone then we have $\mathrm{P}\left(\tau_0(S^{*k}) = 1 | Z_0 = z\right) = 1$ for all $z \in S^{*k}$ and hence $\tau_0(S^{*k})$ is optimal (given conditions (7.2.1) and (7.4.6)).

11 Example: No-Information Version of the Best Choice Problem with Random Population Size

In this chapter we consider the best choice problem, also called secretary, beauty contest, dowry or marriage problem. We consider the no-information version with a random number of observations. Given these constraints the problem was posed first by Presman and Sonin, see [PS73]. Here we use a setting based on and compatible with that in [Irl80, part 1]. Neither recall nor uncertainty of selection is allowed, one choice must be made. The relative ranks of the items are observed sequentially with the object of selecting the absolut best one.

First we give the setting in Part 11.1 (page 101) and Part 11.2 (page 102). In [Irl80, part 3] we find some examples of how the FII Algorithm works in this setting. *We found out that these examples have a common structure, which is general and independent of the examples.*

We will show in Theorem 11.13 (page 109) that for each $k \in \mathbb{N}_0$ the $C^{(k)} \in \mathcal{P}ot(\Omega)^{\mathbb{N}}$ of the algorithm can be expressed by some $B^{(k)} \subseteq \mathbb{N}$ (when starting with $C^{(0)} :\equiv \Omega$), a noteworthy simplification.

We will further show (in Lemma 11.10 (page 106)) that the algorithmic step of going from some C to $C^{*\kappa}$ by calculating certain conditional expectations (as defined in Lemma 5.11 (page 55)) can be done by going from B to $B^{*\kappa}$ by direct evaluation of a function given in an analytic form (defined in Setting 11.3 (page 102)).

11.1 General Setting

Consider some $p \in [0,1]^{\mathbb{N}}$ with $\sum_{i=1}^{\infty} p_i = 1$.
For all $k \in \mathbb{N}$ define
$$\Omega_k := \{\sigma : \mathbb{N}_{\leqslant k} \longrightarrow \mathbb{N}_{\leqslant k} \; ; \; \sigma \text{ bijective}\}.$$
Define $\Omega := \bigcup_{k \in \mathbb{N}} \Omega_k$, $\mathcal{A} := \mathcal{P}ot(\Omega)$ and $P : \mathcal{A} \longrightarrow [0,1]$ by
$$P(\{\omega\}) = \frac{p_k}{k!}$$
for all $\omega \in \Omega_k$ and $k \in \mathbb{N}$.
Define $N : \Omega \longrightarrow \mathbb{N}$ as the mapping of each ω onto the unique $k \in \mathbb{N}$ with $\omega \in \Omega_k$.

The maximal possible size of the population is of importance for the following considerations, hence define
$$n_0 := \inf\{n \in \mathbb{N} \; ; \; P(\{N \leqslant n\}) = 1\}$$

and use the convention that the infimum of an empty set is infinity, hence
$$n_0 \in \mathbb{N} \cup \{\infty\}.$$
For all $i \in \mathbb{N}$ define
$$A_i : \Omega \longrightarrow \mathbb{N}, \; \omega \longmapsto \begin{cases} \omega_i & \text{if } k \in \mathbb{N} \text{ exists with } \omega \in \Omega_k \text{ and } i \leq k \\ \infty & \text{otherwise,} \end{cases}$$
and
$$R_i : \Omega \longrightarrow \mathbb{N}, \; \omega \longmapsto \begin{cases} |\{j \in \mathbb{N} \; ; \; j \leq i, \omega_j \leq \omega_i\}| & \text{if } k \in \mathbb{N} \text{ exists with } \omega \in \Omega_k \text{ and } i \leq k \\ \infty & \text{otherwise.} \end{cases}$$

For all $k \in \mathbb{N}$ define $\mathcal{A}_k := \sigma(\{R_i \; ; \; i \in \mathbb{N}_{\leq k}\})$.
For all $k \in \mathbb{N}$ let $q_k : \mathbb{N} \cup \{\infty\} \longrightarrow \mathbb{R}_{\geq 0}$ with $q_k(\infty) = 0$.
For all $k \in \mathbb{N}$ let $X_k := \mathrm{E}\left(q_k(A_k)|\mathcal{A}_k\right)$

N is the number of available elements.
$q_k(i)$ is the payoff you get, if you stop at the kth element and this has the absolut rank i.
R_i is the relative rank and A_i the absolute rank of the ith element.
They are both infinite iff the number of available elements is strictly smaller than i.

11.2 Special Setting

Consider some $d \in]0,1]$.
For all $k \in \mathbb{N}$ and $i \in \mathbb{N} \cup \{\infty\}$ let $q_k(i) := \delta_{i,1} d^k$.

Here the best choice is maximizing the possibility of having the best of all possible candidates.

11.3 Definitions and Remarks

- Define
$$\varphi : \mathbb{N} \times \mathbb{N} \longrightarrow \mathbb{R}_{\geq 0}, \; (n,r) \longmapsto \begin{cases} \dfrac{n}{P(N \geq n)} \sum_{i=n}^{n_0} \dfrac{p_i}{i} & \text{if } P(N \geq n) > 0 \text{ and } r = 1 \\ 0 & \text{otherwise.} \end{cases}$$

We have by Lemma 11.4 (page 104) below
$$X_k = \mathrm{E}\left(q_k(A_k)|\mathcal{A}_k\right) = d^k P(A_k = 1|\mathcal{A}_k) = d^k \varphi(k, R_k) \text{ for all } k \in \mathbb{N}.$$

- Since $\bigcup_{n \in \mathbb{N}} \bigcap_{k \geq n} \{R_k = \infty\} = \Omega$ we have $0 = \lim_{n \to \infty} X_n$ pointwise, so by defining
$$X_\infty := 0$$
we have $X_\infty = \limsup_{n \to \infty, k \geq n} X_k$.
By Lemma 5.7 (page 53) the assumptions (5.2.1) and (5.2.2) are true.

- Define stopping areas for all $B \subseteq \mathbb{N}_{<n_0}$ and $n \in \mathbb{N} \cup \{\infty\}$ by

$$C_n(B) := \begin{cases} \{R_n \in \{1,\infty\}\} & \text{if } n \in B \\ \Omega & \text{if } n \geqslant n_0 \\ \{R_n = \infty\} & \text{otherwise.} \end{cases}$$

In the examples examined in [Irl80, part 3] the appearing stopping areas were all of this kind. We will show in the sequel that this was not due to the examples, it's a general result for this best choice problem.

Define a sequence of stopping areas for all $B \subseteq \mathbb{N}_{<n_0}$ by

$$C(B) := (C_n(B))_{n \in \mathbb{N} \cup \{\infty\}}.$$

Hence we have constructed for each $B \subseteq \mathbb{N}_{<n_0}$ some $C(B) \in \mathcal{C}$. In Theorem 11.13 (page 109) we show that for each $k \in \mathbb{N}$ there is some $B^{(k)} \subseteq \mathbb{N}_{<n_0}$ such that $C^{(k)} = C(B^{(k)})$ (when starting with $C^{(0)} := \Omega$).

- Using the same definitions as in Definition 5.9 (page 54) we have for all $B \subseteq \mathbb{N}_{<n_0}$ and $n \in \mathbb{N}_0$

$$\tau_n(C(B)) = (n \vee n_0) \wedge \inf\{l \in \mathbb{N} \cap [n, n_0[\ ;\ R_l = \infty\} \\ \wedge \inf\{l \in B \cap [n, n_0[\ ;\ R_l = 1\},$$

and thus

$$\tau_1(C(B)) = n_0 \wedge \inf\{l \in \mathbb{N}_{<n_0}\ ;\ R_l = \infty\} \\ \wedge \inf\{l \in B\ ;\ R_l = 1\}.$$

- As mentioned above the algorithmic step can be done by calculating for some B a $B^{*\kappa}$ by using a deterministic function in analytical form as follows.

Define
(using the convention that an empty sum is zero and an empty product is one)

$$a : \mathbb{N}_{<n_0} \times [0,1] \times Pot(\mathbb{N}_{<n_0}) \times \mathbb{N} \longrightarrow \mathbb{R},$$

$$(n, d, B, j) \longmapsto \sum_{l=n}^{n_0} \frac{p_l}{l} \left(nd^n - \sum_{i \in [n+j,l] \cap (B \cup \{n_0\})} d^i \cdot \prod_{k \in B \cap [n+j,i[} \frac{k-1}{k} \right).$$

Hence for all $n \in \mathbb{N}_{<n_0}$ and $j \in \mathbb{N}$ we have

(11.3.1) $\qquad a(n, d, \mathbb{N}_{<n_0}, j) = \sum_{l=n}^{n_0} \frac{p_l}{l} \left(nd^n - (n+j-1) \sum_{i=n+j}^{l} \frac{d^i}{i-1} \right).$

Define[1] further for all $\kappa \in \mathbb{N} \cup \{\infty\}$
(using the convention that the infimum of an empty set is infinity)
$$a_\kappa : \mathbb{N}_{<n_0} \times [0,1] \times Pot(\mathbb{N}_{<n_0}) \longrightarrow \mathbb{R} \cup \{-\infty, \infty\},$$
$$(n, d, B) \longmapsto \inf \{a(n, d, B, j) \,;\, j \in \mathbb{N} \text{ with } j < \kappa + 1\}.$$

Define for all $\kappa \in \mathbb{N} \cup \{\infty\}$ and $B \subseteq \mathbb{N}_{<n_0}$
$$B^{*\kappa} := \{n \in B \,;\, a_\kappa(n, d, B) \geq 0\}.$$

11.4 Lemma

For all $n \in \mathbb{N}$ we have
$$P(A_n = 1 | \mathcal{A}_n) = P(A_n = 1 | R_n = 1) \cdot \mathbf{1}_{\{R_n = 1\}}$$
and if $P(N \geq n) > 0$ we also have
$$P(A_n = 1 | R_n = 1) = \frac{n}{P(N \geq n)} \sum_{i=n}^{n_0} \frac{p_i}{i}.$$

Proof: A well-known result of the theory of best-choice problems. $\square$

11.5 Lemma

Let $n \in \mathbb{N}_{<n_0}$, $B \subseteq \mathbb{N}_{<n_0}$, $j \in \mathbb{N}$. We have
$$\mathrm{E}\left(X_{\tau_{n+j}(C(B))} \big| \mathcal{A}_n\right) = \sum_{\substack{i \in B \cup \{n_0\} \\ i \geq n+j}} d^i \varphi(i, 1) \cdot P(R_i = 1, R_k \neq 1 \text{ for all } k \in B \cap [n+j, i[\, | \mathcal{A}_n).$$

Proof: We have for all $i \in \mathbb{N}$ with $n + j \leq i \leq n_0$
$$\mathrm{E}\left(X_i \mathbf{1}_{\{i = \tau_{n+j}(C(B))\}} \big| \mathcal{A}_n\right)$$
$$= \mathrm{E}\left(d^i \varphi(i, R_i) \mathbf{1}_{\{i = \tau_{n+j}(C(B))\}} \big| \mathcal{A}_n\right)$$
$$= \mathrm{E}\left(d^i \varphi(i, 1) \mathbf{1}_{\{R_i = 1\}} \mathbf{1}_{\{i = \tau_{n+j}(C(B))\}} \big| \mathcal{A}_n\right)$$
$$= d^i \varphi(i, 1) \cdot P(i = \tau_{n+j}(C(B)), R_i = 1 | \mathcal{A}_n)$$
$$= d^i \varphi(i, 1) \cdot P(i = \inf\{p \in B \cup \{n_0\} \,;\, p \geq n+j, R_p = 1\}, R_i = 1 | \mathcal{A}_n)$$
$$= \begin{cases} d^i \varphi(i, 1) \cdot P(R_i = 1, R_k \neq 1 \text{ for all } k \in B \cap [n+j, i[\, | \mathcal{A}_n) & \text{if } i \in B \cup \{n_0\} \\ 0 & \text{otherwise} \end{cases}$$

Hence we have
$$\mathrm{E}\left(X_{\tau_{n+j}(C(B))} \big| \mathcal{A}_n\right)$$
$$= \mathrm{E}\left(\sum_{i=n+j}^{\infty} X_i \mathbf{1}_{\{i = \tau_{n+j}(C(B))\}} \bigg| \mathcal{A}_n\right)$$
$$= \sum_{\substack{i \in B \cup \{n_0\} \\ i \geq n+j}} d^i \varphi(i, 1) \cdot P(R_i = 1, R_k \neq 1 \text{ for all } k \in B \cap [n+j, i[\, | \mathcal{A}_n). \quad \square$$

---

[1] In [Irl80, part 3] this function had been used and defined only for $B = \mathbb{N}_{<n_0}$ and $\kappa = 1$. The expressions $a_1(n, d, \mathbb{N}_{<n_0}, 1)$ and $a(n, d, \mathbb{N}_{<n_0}, 1)$ herein are equal to $n \cdot c_n(d)$ therein.

11.6 Lemma

For all finite sets $I \subseteq \mathbb{N}$ and $b \in \mathbb{N}^I$ with $b_i \leqslant i$ for all $i \in I$ we have

$$P(\forall\; i \in I\;\; R_i = b_i | N \geqslant \max I) = \prod_{i \in I} \frac{1}{i}.$$

11.7 Lemma

For all finite sets $I \subseteq \mathbb{N}$ and $D \in Pot(\mathbb{N})^I$ with $D_i \subseteq \mathbb{N}_{\leqslant i}$ for all $i \in I$ we have

$$P(\forall\; i \in I\;\; R_i \in D_i | N \geqslant \max I) = \prod_{i \in I} \frac{|D_i|}{i}.$$

Proof: By Lemma 11.6 (page 105). $\square$

11.8 Lemma

Let $B \subseteq \mathbb{N}_{<n_0}$, $n, i \in \mathbb{N}$ with $n < i \leqslant n_0$, $j \in \mathbb{N}$. Then we have

$$P(R_i = 1 \wedge \forall\; k \in B \cap [n+j, i[\;\; R_k \neq 1 | R_n = 1) = \frac{P(N \geqslant i)}{i} \frac{1}{P(N \geqslant n)} \prod_{k \in B \cap [n+j, i[} \frac{k-1}{k}.$$

Proof: We have

$$P(R_n = 1) = P(R_n < \infty) \cdot P(R_n = 1 | R_n < \infty) = \frac{1}{n} P(N \geqslant n)$$

and

$$P(R_i = 1 \wedge \forall\; k \in B \cap [n+j, i[\;\; R_k \neq 1 \wedge R_n = 1)$$
$$= P(R_i < \infty) \cdot P(R_i = 1 \wedge R_n = 1 \wedge \forall\; k \in B \cap [n+j, i[\;\; R_k \neq 1 | R_i < \infty)$$
$$\stackrel{11.7}{=} P(R_i < \infty) \cdot \left(\frac{1}{i} \frac{1}{n} \prod_{k \in B \cap [n+j, i[} \frac{k-1}{k} \right)$$
$$= \frac{1}{n} \frac{P(N \geqslant i)}{i} \prod_{k \in B \cap [n+j, i[} \frac{k-1}{k}. \qquad \square$$

11.9 Lemma

Let $B \subseteq \mathbb{N}_{<n_0}$, $n \in B$, $j \in \mathbb{N}$, $\omega \in \{R_n = 1\}$. We have

$$\left(X_n - E\left(X_{\tau_{n+j}(C(B))} | \mathcal{A}_n \right) \right)(\omega) \geqslant 0 \iff a(n, d, B, j) \geqslant 0.$$

Proof: For any sequence $(c_i)_{i\in\mathbb{N}}$ we have

$$\sum_{i\in(B\cup\{n_0\})\cap[n+j,n_0]} d^i\left(\varphi(i,1)\cdot\frac{P(N\geqslant i)}{i}\right)c_i$$

(11.9.1) $$\overset{11.3}{=} \sum_{i\in(B\cup\{n_0\})\cap[n+j,n_0]} d^i\left(\sum_{l=i}^{n_0}\frac{p_l}{l}\right)c_i$$

$$= \sum_{l=n}^{n_0}\frac{p_l}{l}\sum_{i\in[n+j,l]\cap(B\cup\{n_0\})} d^i c_i.$$

Since $n < n_0$ we have $P(N \geqslant n) > 0$ and hence we have

$$\left(X_n - \mathrm{E}\left(X_{\tau_{n+j}(C(B))}|\mathcal{A}_n\right)\right)(\omega)$$

$$\overset{11.5}{=} d^n\varphi(n,1) - \sum_{i\in(B\cup\{n_0\})\cap[n+j,n_0]} d^i\varphi(i,1)\cdot P(R_i = 1, \forall\, k \in B\cap[n+j,i[\ R_k \neq 1|\mathcal{A}_n)(\omega)$$

$$\overset{11.8}{=} d^n\varphi(n,1) - \sum_{i\in(B\cup\{n_0\})\cap[n+j,n_0]} d^i\varphi(i,1)\cdot\frac{P(N\geqslant i)}{i}\cdot\frac{1}{P(N\geqslant n)}\prod_{k\in B\cap[n+j,i[}\frac{k-1}{k}$$

$$\overset{(11.9.1)}{\underset{(11.4)}{=}} \frac{1}{P(N\geqslant n)}\sum_{l=n}^{n_0}\frac{p_l}{l}\left[nd^n - \sum_{i\in[n+j,l]\cap(B\cup\{n_0\})} d^i\prod_{k\in B\cap[n+j,i[}\frac{k-1}{k}\right]$$

$$\overset{(11.3)}{=} \frac{1}{P(N\geqslant n)}a(n,d,B,j). \qquad \square$$

11.10 Theorem[2]

For all $B \subseteq \mathbb{N}_{<n_0}$, $\kappa \in \mathbb{N}\cup\{\infty\}$ we have

$$(C(B))^{*\kappa} = C(B^{*\kappa}),$$

where the first $^{*\kappa}$ is defined in Definition 5.9 (page 54), while the latter $^{*\kappa}$ as well as the mapping C is defined in Setting 11.3 (page 102).

Proof: Let $B \subseteq \mathbb{N}_{<n_0}$ and $n \in \mathbb{N}$.

On $\{R_{n+1} = \infty\}$ we have $X_k = 0$ for all $k \in \mathbb{N}$ with $k > n$, hence we have $\tau_{n+j}(C(B)) = n + j > n$ for all $j \in \mathbb{N}$ and thus

$$X_{\tau_{n+j}(C(B))} = 0 \text{ and } X_n - \mathrm{E}\left(X_{\tau_{n+j}(C(B))}|\mathcal{A}_n\right) \geqslant 0 \text{ for all } j \in \mathbb{N}.$$

This shows

(11.10.1) $$\{R_n = \infty\} \subseteq \{R_{n+1} = \infty\} \subseteq (C(B))_n^{*\kappa}.$$

<u>Case of $n \geqslant n_0$:</u>
We have $C_n(B) = \Omega$ and $\{R_{n+1} = \infty\} = \Omega$. Hence

$$(C(B))_n^{*\kappa} = \Omega.$$

---

[2] For $\kappa = 1$ and $B = \mathbb{N}_{<n_0}$ this can be found in [Irl80, 3.2] with $n \cdot c_n(d) = a(n, d, \mathbb{N}_{<n_0}, 1) = a(n, d, \mathbb{N}_{<n_0})$.

Case of $n \in B$:
We have $C_n(B) = \{R_n \in \{1, \infty\}\}$.
For all $\omega \in \{R_n = 1\}$ we have by Lemma 11.9 (page 105) the following equivalence:

$$\omega \in (C(B))_n^{*\kappa}$$
$$\stackrel{\text{def.}}{\iff} \left(X_n - \operatorname{E}\left(X_{\tau_{n+j}(C(B))}\big|\mathcal{A}_n\right)\right)(\omega) \geq 0 \text{ for all } j \in \mathbb{N} \text{ with } j < \kappa + 1$$
$$\stackrel{11.9}{\iff} a(n, d, B, j) \geq 0 \text{ for all } j \in \mathbb{N} \text{ with } j < \kappa + 1$$
$$\stackrel{\text{by def.}}{\iff} a_\kappa(n, d, B) \geq 0$$
$$\stackrel{\text{def.}}{\iff} n \in B^{*\kappa}.$$

Hence we have
$$\{R_n = 1\} \subseteq (C(B))_n^{*\kappa} \iff n \in B^{*\kappa}$$
and
$$\{R_n = \infty\} \stackrel{(11.10.1)}{\subseteq} (C(B))_n^{*\kappa} \stackrel{\text{by def.}}{\subseteq} C_n(B) \stackrel{\text{def.}}{=} \{R_n \in \{1, \infty\}\}.$$

Hence we have by
$$(C(B))_n^{*\kappa} = \begin{cases} \{R_n \in \{1, \infty\}\} & \text{if } n \in B^{*\kappa} \\ \{R_n = \infty\} & \text{otherwise.} \end{cases}$$

Case of $n < n_0$ and $n \notin B$:
We have
$$C_n(B) \stackrel{\text{def.}}{=} \{R_n = \infty\} \stackrel{(11.10.1)}{\subseteq} (C(B))_n^{*\kappa} \stackrel{\text{by def.}}{\subseteq} C_n(B),$$

hence
$$(C(B))_n^{*\kappa} = \{R_n = \infty\}. \qquad \square$$

11.11 Theorem[3]

Assume $C_n^{(0)} := \Omega$ and $\tau_n^{(0)} :\equiv n$ for all $n \in \mathbb{N} \cup \{\infty\}$.
Then we have
$$C^{(1)} = C(\mathbb{N}_{<n_0}),$$
where the left-hand side is defined in Algorithm 2.13 (page 22) and the mapping C on the right in Setting 11.3 (page 102).

Proof: It is enough to show that

- for all $n \in \mathbb{N}_{<n_0}$ we have $C_n^{(1)} = \{R_n \in \{1, \infty\}\}$ and
- for all $n \in \mathbb{N}_{\geq n_0}$ we have $C_n^{(1)} = \Omega$.

Define $B := \mathbb{N}_{<n_0}$. First let $n \in \mathbb{N}_0$ with $n_0 \leq n$.
Since $R_k \equiv \infty$ for all $k \in \mathbb{N}_0$ with $n_0 + 1 \leq k$ follows

$$\operatorname{E}\left(X_{\tau_{n+j}^{(0)}}\big|\mathcal{A}_n\right) = \operatorname{E}\left(X_{n+j}\big|\mathcal{A}_n\right) = 0 \text{ for all } j \in \mathbb{N} \text{ with } j < \kappa(1) + 1,$$

---
[3]For $\kappa = 1$ this can be found in [Irl80, 3.1].

hence
$$X_n - \mathrm{E}\left(X_{\tau_{n+j}^{(0)}}\Big|\mathcal{A}_n\right) = X_n \geq 0 \text{ for all } j \in \mathbb{N} \text{ with } j < \kappa(1) + 1.$$
Thus we have $C_n^{(1)} = \Omega$ for all $n \in \mathbb{N}_0$ with $n_0 \leq n$.

Next let $n \in \mathbb{N}$ with $n < n_0$.
We have $P(N \geq n) > 0$.
On $\{R_n = \infty\}$ we have $R_k = \infty$ for all $k \in \mathbb{N}_0$ with $n \leq k$.
Hence
$$X_n - \mathrm{E}\left(X_{\tau_{n+j}^{(0)}}\Big|\mathcal{A}_n\right) \geq 0 \text{ for all } j \in \mathbb{N} \text{ with } j < \kappa(k) + 1.$$
Thus we have $\{R_n = \infty\} \subseteq C_n^{(1)}$.

Now consider some $\omega \in \{R_n < \infty\}$ and some $j \in \mathbb{N}$ with $j < \kappa(k) + 1$.
If $n + j > n_0$, then we have $\mathrm{E}(X_{n+j}|\mathcal{A}_n)(\omega) = 0$, hence $(X_n - \mathrm{E}(X_{n+j}|\mathcal{A}_n))(\omega) \geq 0$.
So assume $n + j \leq n_0$. Then we have
$$\begin{aligned}
&\mathrm{E}(\varphi(n+j, R_{n+j})|\mathcal{A}_n)(\omega) \\
&= \varphi(n+j, 1)P(R_{n+j} = 1|\mathcal{A}_n)(\omega) \\
&= \varphi(n+j, 1)P(R_{n+j} = 1|R_n < \infty) \\
&= \varphi(n+j, 1)P(R_{n+j} = 1|R_{n+j} < \infty)\frac{P(R_{n+j} < \infty)}{P(R_n < \infty)} \\
&= \left[\frac{n+j}{P(N \geq n+j)}\sum_{i=n+j}^{n_0}\frac{p_i}{i}\right]\frac{1}{n+j}\frac{P(N \geq n+j)}{P(N \geq n)} \\
&= \frac{1}{P(N \geq n)}\sum_{i=n+j}^{n_0}\frac{p_i}{i},
\end{aligned}$$
hence
$$\begin{aligned}
\mathrm{E}(X_{n+j}|\mathcal{A}_n)(\omega) &= d^{n+j}\mathrm{E}(\varphi(n+j, R_{n+j})|\mathcal{A}_n)(\omega) \\
&= \frac{1}{P(N \geq n)}d^{n+j}\sum_{i=n+j}^{n_0}\frac{p_i}{i} \\
&= \frac{d^n}{P(N \geq n)}d^j\sum_{i=n+j}^{n_0}\frac{p_i}{i}.
\end{aligned}$$
(11.11.1)

If $\omega \in \{R_n \neq 1\} \cap \{R_n < \infty\}$ we have $X_n(\omega) \stackrel{11.3}{=} d^n\varphi(n, R_n)(\omega) = 0$ and $\mathrm{E}(X_{n+j}|\mathcal{A}_n)(\omega) > 0$, hence
$$\left(X_n - \mathrm{E}\left(X_{\tau_{n+j}^{(0)}}\Big|\mathcal{A}_n\right)\right)(\omega) = (X_n - \mathrm{E}(X_{n+j}|\mathcal{A}_n))(\omega) < 0.$$
On $\{R_n = 1\}$ we have
$$\begin{aligned}
X_n &\stackrel{11.3}{=} d^n\varphi(n, 1) \\
&= \frac{nd^n}{P(N \geq n)}\left(\sum_{i=n}^{n+j-1}\frac{p_i}{i} + \sum_{i=n+j}^{n_0}\frac{p_i}{i}\right) \\
&= \frac{d^n}{P(N \geq n)}\left(n\sum_{i=n}^{n+j-1}\frac{p_i}{i} + n\sum_{i=n+j}^{n_0}\frac{p_i}{i}\right)
\end{aligned}$$

and hence

$$\frac{P(N \geq n)}{d^n} \left(X_n - \mathrm{E}\left(X_{n+j}|\mathcal{A}_n\right)\right) \stackrel{11.11.1}{=} n \sum_{i=n}^{n+j-1} \frac{p_i}{i} + \underbrace{(n-d^j)}_{\geq n-1 \geq 0} \cdot \sum_{i=n+1+j}^{n_0} \frac{p_i}{i} \geq 0.$$

Thus it follows

$$\left(X_n - \mathrm{E}\left(X_{\tau_{n+j}^{(0)}}\Big|\mathcal{A}_n\right)\right) > 0 \text{ on } \{R_n = 1\}.$$

So we infer $C_n^{(1)} = \{R_n \in \{1, \infty\}\}$ for all $n \in \mathbb{N}$ with $n < n_0$.

Hence we have $C^{(1)} = C(\mathbb{N}_{<n_0})$. □

11.12 Setting Continued

Let $\kappa : \mathbb{N} \longrightarrow \mathbb{N} \cup \{\infty\}$. Assume $C_n^{(0)} := \Omega$ and $\tau_n^{(0)} :\equiv n$ for all $n \in \mathbb{N} \cup \{\infty\}$.
Define $B^1 := \mathbb{N}_{<n_0}$ and for all $k \in \mathbb{N}$ inductively with * defined in Setting 11.3 (page 102)

$$B^{k+1} = (B^k)^{*\kappa(k)}.$$

Define further

$$B^\infty := \bigcap_{k \in \mathbb{N}} B^k.$$

11.13 Theorem

For all $k \in \mathbb{N}$ we have

$$C^{(k)} = C(B^k),$$

where the left-hand side is defined in Algorithm 2.13 (page 22) and the mapping C on the right in Setting 11.3 (page 102).

$$\tau(C(B^\infty)) \text{ is optimal.}$$

Proof: We start the algorithm with $C^{(0)} \equiv \Omega$, but there is (except for the trivial case of $n_0 = 1$) no $A \subseteq \mathbb{N}$ with $C(A) \equiv \Omega$ (see Definition 11.3 (page 102)). Hence the induction cannot start with $k = 0$. The induction start with $k = 1$ is done in Theorem 11.11 (page 107). Thereby we have $C^{(1)} = C(\mathbb{N}_{<n_0}) = C(B^1)$. For the induction step[4] consider $k \in \mathbb{N}$ with $C^{(k)} = C(B^k)$, then we have

$$C^{(k+1)} \stackrel{\text{def}}{=} (C^{(k)})^{*\kappa(k)} \stackrel{\text{I.V.}}{=} (C(B^k))^{*\kappa(k)} \stackrel{11.10}{=} C((B^k)^{*\kappa(k)}) \stackrel{\text{def}}{=} C(B^{k+1}).$$

The second statement follows by the first with Remark 5.15 (page 64). □

---

[4] The general induction step from $k = 1$ to $k = 2$ can be found in [Irl80, 3.2], therein with $n \cdot c_n(d) = a(n, d, 1, \mathbb{N}_{<n_0})$.
Further induction steps are done in [Irl80] for several examples individually.

11.14 Lemma[5]

For all $l \in \mathbb{N}_{<n_0}$, $B \subseteq \mathbb{N}_{<n_0}$, $\kappa_1, \kappa_2 \in \mathbb{N} \cup \{\infty\}$ with $\kappa_2 \leqslant \kappa_1$ we have

$$\mathbb{N} \cap [l, n_0[\subseteq B^{*\kappa_1} \implies \mathbb{N} \cap [l, n_0[\subseteq (B^{*\kappa_1})^{*\kappa_2}.$$

Proof: By Definition 11.3 (page 102) we have obviously for all $n \in \mathbb{N}_{<n_0}$, $B \subseteq \mathbb{N}_{<n_0}$, $j \in \mathbb{N}$

$$a(n, d, B, j) = a(n, d, B \cap [n+j, n_0[, j)$$

and hence for all $\kappa \in \mathbb{N} \cup \{\infty\}$

$$a_\kappa(n, d, B) = a_\kappa(n, d, B \cap [n, n_0[).$$

Let $l \in \mathbb{N}_{<n_0}$, $B \subseteq \mathbb{N}_{<n_0}$, $\kappa_1, \kappa_2 \in \mathbb{N} \cup \{\infty\}$ with $\mathbb{N} \cap [l, n_0[\subseteq B^{*\kappa_1}$. Let $n \in \mathbb{N} \cap [l, n_0[$. Then we have $B \cap [n, n_0[= \mathbb{N} \cap [n, n_0[= B^{*\kappa_1} \cap [n, n_0[$. Since $n \in B^{*\kappa_1}$ we have

$$0 \leqslant a_{\kappa_1}(n, d, B) \leqslant a_{\kappa_2}(n, d, B) = a_{\kappa_2}(n, d, B^{*\kappa_1}),$$

hence we have $n \in (B^{*\kappa_1})^{*\kappa_2}$. □

11.15 Lemma: Optimality[6]

Consider $B \subseteq \mathbb{N}_{<n_0}$, $k \in \mathbb{N}_{\geqslant 2}$ and some $l \in \mathbb{N}$ with $B^k = \mathbb{N} \cap [l, n_0[$. Then we have

$$B^k = B^{k+1}.$$

Proof: By Lemma 11.14 (page 110) we have

$$B^k \stackrel{11.12}{=} (B^{k-1})^{*\kappa(k)} \subseteq ((B^{k-1})^{*\kappa(k)})^{*\kappa(k)}.$$

The other inclusion is always true by definition. Using Lemma 11.14 (page 110), Theorem 11.13 (page 109), Conclusion 5.13 (page 63) and Theorem 5.12 (page 61) we infer $\tau(C(B^k))$ is optimal. Thus it follows by Lemma 5.10 (page 55) that $C^{(k)} = C^{(k+1)}$ and hence $B^k = B^{k+1}$. □

---

[5]Generalisation of the ideas of [Irl80, Lemma 3.4], therein $B = \mathbb{N}_{<n_0}$.
[6]For $B = \mathbb{N}_{<n_0}$ this can be found in [Irl80, Theorem 3.4].

11.16 Lemma: Problem-Independent Positive Values of a[7]

Let $n \in \mathbb{N}_{<n_0}$, $B \subseteq \mathbb{N}_{<n_0}$.

(11.16.1)   For all $j \in \mathbb{N}$ with $n_0 - 2n \leq j \vee n(1-d) \geq d^j$ we have $a(n, d, B, j) \geq 0$.

Consider $2 \leq n_0 < \infty$.
Then we have

(11.16.2) $$a(n, d, B, n_0 - n - 1) \geq 0$$

and for all $\kappa \in \mathbb{N} \cup \{\infty\}$

(11.16.3) $$a_\kappa(n_0 - 1, d, B) \geq 0.$$

Proof: To prove (11.16.1) let $j \in \mathbb{N}$.

- If $n(1-d) \geq d^j$, then we have $d \neq 1$ and

$$\sum_{i \in [n+j, l] \cap (B \cup \{n_0\})} d^i \cdot \prod_{k \in B \cap [n+j, i[} \frac{k-1}{k} \leq \sum_{i=n+j}^{l} d^i \cdot 1$$

$$\leq d^{n+j} \sum_{i=0}^{\infty} d^i = d^n d^j \frac{1}{1-d} \leq n d^n.$$

- If $n_0 - n \leq j$, then

$$a(n, d, B, j) = \sum_{l=n}^{n_0} \frac{p_l}{l}(nd^n - 0) = nd^n \sum_{l=n}^{n_0} \frac{p_l}{l} \geq 0.$$

- If $n_0 - 2n \leq j < n_0 - n$, then we have

$$n \geq n_0 - n - j = \sum_{i=n+j}^{n_0} 1 \geq \sum_{i=n+j}^{n_0} d^{i-n} = \frac{1}{d^n} \sum_{i=n+j}^{n_0} d^i,$$

hence for all $l \in \mathbb{N}$ with $n \leq l \leq n_0$

$$\sum_{i \in [n+j, l] \cap (B \cup \{n_0\})} d^i \cdot \prod_{k \in B \cap [n+j, i[} \frac{k-1}{k} \leq \sum_{i=n+j}^{n_0} d^i \cdot 1 \leq nd^n.$$

Now consider $2 \leq n_0 < \infty$.
Then we have $-n \leq -1$, hence $n_0 - 2n \leq n_0 - n - 1$.
And thus follows (11.16.2) by (11.16.1).

For all $j \in \mathbb{N}$ we have $n_0 - 2(n_0 - 1) = 2 - n_0 \leq 0 \leq j$,
hence by (11.16.1) $a(n_0 - 1, d, B, j) \geq 0$.
Consider $\kappa \in \mathbb{N} \cup \{\infty\}$. By the definition of a_κ we have (11.16.3). □

---
[7] This is a generalisation of the text behind the proof of [Irl80, 3.6], therein only for $j = 1$, and the ideas of [Irl80, Lemma 3.3], therein only for $B = \mathbb{N}_{<n_0}$.

11.17 Conclusion about a and a_κ

Let $B \subseteq \mathbb{N}_{n_0}$, $n \in B$, $\kappa \in \mathbb{N} \cup \{\infty\}$. We have

$$a_\kappa(n, d, B) \geq 0$$
$$\iff$$
$a(n, d, B, j) \geq 0$ for all $j \in \mathbb{N}$ with $j < \kappa \wedge (n_0 - 2n)$ and $n(1-d) < d^j$.

Proof: This is a consequence of Lemma 11.16 (page 111). □

11.18 Remark

Let $n \in \mathbb{N}$, $j \in \mathbb{N}$ with $n(1-d) \geq d^j$. Then we have

$$\forall\, n' \in \mathbb{N}, j \in \mathbb{N}_0 \ \ n \leq n', j \leq j' \implies n'(1-d) \geq d^{j'}.$$

11.19 Lemma[8]

Let $d < 1$, $B \subseteq \mathbb{N}_{<n_0}$, $\kappa \in \mathbb{N} \cup \{\infty\}$. We have

$$B \cap \left[\frac{d}{1-d}, n_0\right[\subseteq B^{*\kappa}.$$

Proof: Let $n \in B$ with $\frac{d}{1-d} \leq n < n_0$. Then for all $j \in \mathbb{N}$ we have $(1-d)n \geq d^1 \geq d^j$, hence $a(n, d, B, j) \geq 0$. Thus we have $a_\kappa(n, d, B) \geq 0$, hence $n \in B^{*\kappa}$. □

11.20 Conclusion: Reasonable Minimal Value for d

If $d \leq \frac{1}{2}$ we have $B = B^{*\kappa}$ for all $B \subseteq \mathbb{N}_{<n_0}$, $\kappa \in \mathbb{N} \cup \{\infty\}$.

Proof: Let $d \leq \frac{1}{2}$. Then $2d \leq 1$, hence $d \leq 1 - d$, $\frac{d}{1-d} \leq 1$ and thus $\mathbb{N}_{<n_0} \cap [\frac{d}{1-d}, n_0[= \mathbb{N}_{<n_0}$. The rest follows by Lemma 11.19 (page 112). □

11.21 Lemma: Bounded Values

Let $n \in \mathbb{N}_{<n_0}$, $B \subseteq \mathbb{N}_{<n_0}$, $j \in \mathbb{N}$. Then we have

$$|a(n, d, B, j)| \leq \sum_{l=n}^{n_0} p_l \leq 1.$$

Proof: For all $i \in [n+j, n_0]$ we have $d^i \leq d^n$ since $d \leq 1$. Hence we have

$$(11.21.1) \qquad \sum_{i \in [n+j, l] \cap (B \cup \{n_0\})} d^i \cdot \prod_{k \in B \cap [n+j, i[} \frac{k-1}{k} \leq \sum_{i=n+j}^{l} d^i \leq \sum_{i=n+j}^{l} d^n \leq l \cdot d^n.$$

---
[8]Generalisation of [Irl80, text behind the proof of 3.6], therein $j = 1$

Define for all $l \in [n, n_0] \cap \mathbb{N}$

$$c_l := \frac{1}{l}\left(nd^n - \sum_{i \in [n+j,l] \cap (B \cup \{n_0\})} d^i \cdot \prod_{k \in B \cap [n+j,l[} \frac{k-1}{k}\right).$$

Obviously we have

$$-1 \leqslant -d^n = \frac{1}{l}(-l \cdot d^n \cdot 1) \stackrel{(11.21.1)}{\leqslant} c_l \leqslant \frac{1}{l}nd^n \leqslant d^n \leqslant 1,$$

hence $|c_l| \leqslant 1$. So it follows that

$$|a(n,d,B,j)| = \left|\sum_{l=n}^{n_0} p_l c_l\right| \leqslant \sum_{l=n}^{n_0} p_l \, |c_l| \leqslant \sum_{l=n}^{n_0} p_l \leqslant \sum_{l=1}^{n_0} p_l = 1. \qquad \square$$

11.22 Conclusion: Bounded Values

The image of the function a is a subset of $[-1, 1]$
and hence for all $\kappa \in \mathbb{N} \cup \{\infty\}$ the image of a_κ is a subset of $[-1, 1] \cup \{\infty\}$.

Proof: This is a consequence of Lemma 11.21 (page 112). $\qquad \square$

11.23 Lemma: Limit Value Zero[9]

Consider $n_0 = \infty$. Let $B \subseteq \mathbb{N}_{<n_0}$. We have

$$\lim_{n \to \infty} a(n, d, B, j) = 0 \text{ for all } j \in \mathbb{N}$$

and hence for all $\kappa : \mathbb{N} \longrightarrow \mathbb{N} \cup \{\infty\}$

$$\lim_{n \to \infty} a_{\kappa(n)}(n, d, B) = 0.$$

Proof: Since $\lim\limits_{n \to \infty} \sum\limits_{l=n}^{\infty} p_l = 0$, this follows by Lemma 11.21 (page 112). $\qquad \square$

11.24 Theorem[10]

Let $B \subseteq \mathbb{N}_{<n_0}$, $\kappa \in \mathbb{N} \cup \{\infty\}$.

Assume for all $n \in \mathbb{N}_{<n_0-2}$ with $n(1-d) < d$

$$\frac{a_\kappa(n,d,B)}{n} \geqslant \frac{a_\kappa(n+1,d,B)}{n+1} \implies \frac{a_\kappa(n+1,d,B)}{n+1} \geqslant \frac{a_\kappa(n+2,d,B)}{n+2}.$$

Then we have

(11.24.1) $$B^{*\kappa} = \emptyset$$

or

(11.24.2) $$\text{there is } l \in B \text{ with } B^{*\kappa} = B \cap [l, n_0[.$$

---
[9] Generalisation of [Irl80, Remark behind Lemma 3.3], therein for $B = \mathbb{N}_{<n_0}$
[10] Generalisation of [Irl80, Proposition 3.6.a], therein for $B = \mathbb{N}_{<n_0}$.

Proof: Since this is obvious for $n_0 \leq 2$, consider $n_0 > 2$.
We will prove now

(11.24.3) $\qquad \forall\, p \in B \backslash B^{*\kappa} \ \ \forall\, k \in B^{*\kappa} \ \ k \notin]p, n_0[.$

Suppose $p \in B \backslash B^{*\kappa}$. Since $p \notin B^{*\kappa}$, we have $a_\kappa(p, d, B) < 0$.
Assume the existence of some $k \in B^{*\kappa}$ with $k < p$, hence $k \in B$ and $a_\kappa(k, d, B) \geq 0$.
Then there is $m \in \mathbb{N}$ with $k \leq m \leq p$ and $a_\kappa(m, d, B) \geq 0 > a_\kappa(m+1, d, B)$.

If $2 < n_0 < \infty$, it follows by induction over n that $a_\kappa(n_0 - 1, d, B) < 0$, contradicting (11.16.3).
If $n_0 = \infty$, it follows that $\lim_{n \to \infty} a_\kappa(n, d, B) < 0$, contradicting Lemma 11.23 (page 113).

Now consider the situation of $B^{*\kappa} \neq \emptyset$ and let l be the minimal element of $B^{*\kappa}$. Assume the existence of some $p \in B \backslash B^{*\kappa}$ with $p > l$, then we have by (11.24.3) that $B^{*\kappa} \subseteq]p, n_0[$ which is a contradiction to the minimality of l. Hence we have $B^{*\kappa} = B \cap [l, n_0[$. $\qquad \square$

11.25 Theorem[11]

For all $k \in \mathbb{N}$ we have $B^k \neq \emptyset$.

Proof: We will show this by induction.
By definition we have $B^1 = \mathbb{N}_{<n_0} \neq \emptyset$.

Let us first consider the case $n_0 < \infty$. By equation (11.16.3) we have $a_{\kappa(k)}(n_0 - 1, d, B^k) \geq 0$ for all $k \in \mathbb{N}$, hence it follows (due to $n_0 - 1 \in \mathbb{N}_{<n_0} = B^1$) inductively that $n_0 - 1 \in B^k$ for all $k \in \mathbb{N}$.

Now let us consider the case $n_0 = \infty$. Let $k \in \mathbb{N}$ with $B^k \neq \emptyset$. Assume for all $n \in B^k$ that $a_{\kappa(k)}(n, d, B^k) < 0$. Then we have $\emptyset = B^{k+1}$, hence $B^{k+2} = \emptyset$ and thus $B^{k+1} = B^{k+2}$. The assumptions made in Conclusion 5.13 (page 63) and Theorem 5.12 (page 61) are obviously fulfilled.[12] By the mentioned theorem and conlusion and Lemma 11.10 (page 106) $\inf\{p \in \mathbb{N}\,;\, R_p = \infty\} = \tau_0(C(\emptyset))$ is optimal with $E\left(X_{\tau_0(C(\emptyset))}\right) = 0$. We have $E(X_1) > 0$, a contradiction. Hence $B^{k+1} \neq \emptyset$. $\qquad \square$

11.26 Lemma

Let $n \in \mathbb{N}_{<n_0}$ and $B \subseteq \mathbb{N}_{<n_0}$ with $n + 1 \in B \cup \{n_0\}$. Then we have

$$\frac{a(n, d, B, 1)}{n} - \frac{a(n+1, d, B, 1)}{n+1} = \frac{d^n}{n}\left(p_n + (n(1-d) - d)\sum_{l=n+1}^{n_0} \frac{p_l}{l}\right),$$

hence

$$\frac{a(n, d, B, 1)}{n} \geq \frac{a(n+1, d, B, 1)}{n+1} \quad \text{iff} \quad p_n \geq (d - n(1-d))\sum_{l=n+1}^{n_0} \frac{p_l}{l}.$$

---

[11]Generalisation of [Irl80, Lemma 3.3], therein $k = 2$ and $\kappa = 1$, hence it was shown that $(\mathbb{N}_{<n_0})^* \neq \emptyset$.
[12]See Setting 11.3 (page 102) and [Irl80, part 3].

Proof: We have

$$\frac{a(n,d,B,1)}{n} = \frac{1}{n}\sum_{l=n}^{n_0}\frac{p_l}{l}\left(nd^n - \sum_{i\in[n+1,l]\cap(B\cup\{n_0\})}d^i\cdot\prod_{k\in B\cap[n+1,i[}\frac{k-1}{k}\right)$$

$$= \frac{d^n}{n}\sum_{l=n}^{n_0}\frac{p_l}{l}\left(n - \sum_{i\in[n+1,l]\cap(B\cup\{n_0\})}d^{i-n}\cdot\prod_{k\in B\cap[n+1,i[}\frac{k-1}{k}\right)$$

$$= \frac{d^n}{n}\left(p_n + \sum_{l=n+1}^{n_0}\frac{p_l}{l}\left(n - d - \underbrace{\sum_{i\in[n+2,l]\cap(B\cup\{n_0\})}d^{i-n}\cdot\frac{n}{n+1}\prod_{k\in B\cap[n+2,i[}\frac{k-1}{k}}_{C:=}\right)\right)$$

and

$$\frac{a(n+1,d,B,0)}{n+1} = \frac{1}{n+1}\sum_{l=n+1}^{n_0}\frac{p_l}{l}\left((n+1)d^{n+1} - \sum_{i\in[n+2,l]\cap(B\cup\{n_0\})}d^i\cdot\prod_{k\in B\cap[n+2,i[}\frac{k-1}{k}\right)$$

$$= \frac{d^n}{n}\left(\sum_{l=n+1}^{n_0}\frac{p_l}{l}\left(nd - \underbrace{\frac{n}{n+1}\sum_{i\in[n+2,l]\cap(B\cup\{n_0\})}d^{i-n}\cdot\prod_{k\in B\cap[n+2,i[}\frac{k-1}{k}}_{C=}\right)\right).$$

Hence it follows

$$\frac{a(n,d,B,0)}{n} - \frac{a(n+1,d,B,0)}{n+1} = \frac{d^n}{n}\left(p_n + \sum_{l=n+1}^{n_0}\frac{p_l}{l}(n-d-C-(nd-C))\right)$$

$$= \frac{d^n}{n}\left(p_n + (n(1-d)-d)\sum_{l=n+1}^{n_0}\frac{p_l}{l}\right). \qquad \square$$

11.27 Theorem: Deterministic Population Size[13]

If $n_0 < \infty$ and $p_{n_0} = 1$, then

there is $l \in \mathbb{N}_{<n_0}$ with $B^2 = B^\infty = \mathbb{N}\cap[l,n_0[$.

Proof: By Lemma 11.15 (page 110) it is enough to show that there is $l \in \mathbb{N}$ with

$$(\mathbb{N}_{<n_0})^{*\kappa(1)} = \mathbb{N}\cap[l,n_0[.$$

Therefore it suffices to show that we have

$$a_{\kappa(1)}(n,d,\mathbb{N}_{<n_0}) \geq 0 \implies a_{\kappa(1)}(n+1,d,\mathbb{N}_{<n_0}) \geq 0 \quad \text{for all } n \in \mathbb{N} \text{ with } n+1 < n_0.$$

---

[13]See also [Irl80, 3.5].

For all $n \in \mathbb{N}_{<n_0}$ and $j \in \mathbb{N}$ we have

$$a(n, d, \mathbb{N}_{<n_0}, j) \stackrel{(11.3.1)}{=} \sum_{l=n}^{n_0} \frac{p_l}{l}\left(nd^n - (n+j-1)\sum_{i=n+j}^{l} \frac{d^i}{i-1}\right)$$

$$= \frac{p_{n_0}}{n_0}(n+j-1)\left(\frac{n}{n+j-1}d^n - \sum_{i=n+j}^{n_0} \frac{d^i}{i-1}\right)$$

and hence

(11.27.1) $\quad a(n, d, \mathbb{N}_{<n_0}, j) \geq 0 \iff \sum_{i=n+j}^{n_0} \frac{d^i}{i-1} \leq \frac{n}{n+j-1}d^n.$

Let $n \in \mathbb{N}$ with $n+1 < n_0$ and assume $a_{\kappa(1)}(n, d, \mathbb{N}_{<n_0}) \geq 0$.

Let $j \in \mathbb{N}$ with $j < n_0 - (n+1)$ and $n(1-d) < d^j$. Then we have

(11.27.2) $\quad n - d^j < n - n(1-d) = nd$

and by Conclusion 11.17 (page 112)

(11.27.3) $\quad a(n, d, \mathbb{N}_{<n_0}, j) \geq 0.$

Then we have

$$\sum_{i=(n+1)+j}^{n_0} \frac{d^i}{i-1} = \sum_{i=n+j}^{n_0} \frac{d^i}{i-1} - \frac{d^{n+j}}{n+j-1}$$

$$\stackrel{(11.27.3)}{\underset{(11.27.1)}{\leq}} \frac{n}{n+j-1}d^n - \frac{d^{n+j}}{n+j-1}$$

$$= \frac{n-d^j}{n+j-1}d^n$$

$$\stackrel{(11.27.2)}{<} \frac{nd}{n+j-1}d^n < \frac{(n+1)}{(n+1)+j-1}d^{n+1}.$$

Hence by (11.27.1) we have $a(n+1, d, \mathbb{N}_{<n_0}, j) \geq 0$.

By Conclusion 11.17 (page 112) we have

$$a_{\kappa(1)}(n+1, d, \mathbb{N}_{<n_0}) \geq 0. \qquad \square$$

11.28 Conclusion[14]

If

(11.28.1) $$\frac{p_{n+1} \cdot p_{n+k}}{n+k} \geq \frac{p_n \cdot p_{n+k+1}}{n+k+1} \text{ for all } k \in \mathbb{N}_{<n_0-n} \text{ and } n \in \mathbb{N}_{<n_0-2}$$

or

(11.28.2) $$p_n \leq p_{n+1} \text{ for all } n \in \mathbb{N}_{<n_0-2},$$

then $B^2 = B^\infty$ and hence $C^{(2)} = C^{(\infty)}$.

Proof: If we have (11.28.2), then Lemma 11.26 (page 114) easily implies the assertion of Theorem 11.24 (page 113) for $B = B^1 = \mathbb{N}_{<n_0}$. Next we will show that (11.28.1) also implies that assertion.

Hence assume (11.28.1) and let $n \in \mathbb{N}_{<n_0-2}$ with $n(1-d) < d$.

If $p_n = 0$ and $\frac{a(n,d,B,0)}{n} \geq \frac{a(n+1,d,B,0)}{n+1}$, then

$$p_{n+1} \geq 0 = p_n \geq (d - n(1-d)) \sum_{l=n+1}^{n_0} \frac{p_l}{l} \geq (d - n(1-d)) \sum_{l=n+2}^{n_0} \frac{p_l}{l}$$

hence we have $\frac{a(n+1,d,B,0)}{n} \geq \frac{a(n+2,d,B,0)}{n+1}$ by Lemma 11.26 (page 114).

By (11.28.1) with $k = n_0 - n - 1$ and $p_{n+k+1} = p_{n_0} > 0$ we have

$$p_n > 0 \text{ implies } p_{n+1} > 0 .$$

Assume
$$\frac{a(n,d,B,0)}{n} \geq \frac{a(n+1,d,B,0)}{n+1}.$$

Then we have by Lemma 11.26 (page 114)

$$p_n \geq (d - n(1-d)) \sum_{l=n+1}^{n_0} \frac{p_l}{l}.$$

We have

$$\sum_{l=n+1}^{n_0} \frac{p_l}{l \cdot p_n} \geq \sum_{k=1}^{n_0-n} \frac{p_{n+k}}{(n+k)p_n} \stackrel{(11.28.1)}{\geq} \sum_{k=1}^{n_0-n-1} \frac{p_{n+k+1}}{(n+k+1)p_{n+1}} \geq \sum_{l=n+1+1}^{n_0} \frac{p_l}{l \cdot p_{n+1}}$$

and $d - n(1-d) > d - (n+1)(1-d)$. Hence we have

$$1 \geq (d - n(1-d)) \sum_{l=n+1}^{n_0} \frac{p_l}{l} \geq (d - (n+1)(1-d)) \sum_{l=n+1+1}^{n_0} \frac{p_l}{l \cdot p_{n+1}},$$

hence

$$p_{n+1} \geq (d - (n+1)(1-d)) \sum_{l=n+1+1}^{n_0} \frac{p_l}{l}$$

---

[14]See [Irl80, 3.6.b.1]

and thus
$$\frac{a(n+1,d,B,0)}{n+1} \geqslant \frac{a(n+2,d,B,0)}{n+2}.$$
Hence the assertion of Theorem 11.24 (page 113) is fulfilled for $B = B^1 = \mathbb{N}_{<n_0}$.
Since Theorem 11.25 (page 114) says that $B^2 \neq \emptyset$, we do not have (11.24.1).
Hence we have (11.24.2).
We infer by Lemma 11.15 (page 110) that $B^2 = B^3$, hence $B^2 = B^\infty$. □

11.29 Remark[15]

(11.28.1) is fulfilled for geometric, Poisson, binomial, hypergemetric, and Laplace distributions.

11.30 Conclusion[16]

Consider $l, m, s \in \mathbb{N}$ with $l < m < s$ and $B^2 = \{n \in \mathbb{N} \,;\, l < n < m \text{ or } s < n < n_0\}$.

If
$$a_{\kappa(2)}(n,d,B^2) \geqslant 0 \text{ for all } n \in \mathbb{N} \text{ with } l < n < m,$$
we have $B^2 = B^\infty$ and hence $C^{(2)} = C^{(\infty)}$.

If
$$a_{\kappa(2)}(n,d,B^2) < 0 \text{ for all } n \in \mathbb{N} \text{ with } l < n < m,$$
we have $B^3 = B^\infty = \{n \in \mathbb{N} \,;\, s < n < n_0\}$ and hence $C^{(3)} = C^{(\infty)}$.

---

[15]See [Irl80, text between 3.6 and 3.7]
[16]See [Irl80, 3.8]

Bibliography

[Asm03] Søren Asmussen. *Applied Probability and Queues.* Applications of mathematics 51. Springer, 2003.

[BKS06] Christian Bender, Anastasia Kolodko, and John Schoenmakers. Policy iteration for American options: overview. *Monte Carlo Methods Appl.*, 12(5-6):347–362, 2006.

[BKS08] Christian Bender, Anastasia Kolodko, and John Schoenmakers. Enhanced policy iteration for american options via scenario selection. *Quantitative Finance*, 8(2):135–146, 2008.

[BS06] Christian Bender and John Schoenmakers. An iterative method for multiple stopping: convergence and stability. *Advances in Applied Probability*, 38(3):729–749, 2006.

[CRS71] Yuan Shih Chow, Herbert Robbins, and David Siegmund. *Great expectations: the theory of optimal stopping.* Houghton Mifflin Company Boston, 1971.

[Gla03] Paul Glasserman. *Monte Carlo Methods in Financial Engineering (Stochastic Modelling and Applied Probability).* Springer, 2003.

[How60] Ronald A. Howard. *Dynamic Programming and Markov Processes.* MIT Press, Cambridge, MA, 1960.

[Irl79] Albrecht Irle. Monotone stopping problems and continous time process. *Zeitschrift für Wahrscheinlichkeitstheorie und Verwandte Gebiete*, 48:49–56, 1979.

[Irl80] Albrecht Irle. On the best choice problem with random population size. *Zeitschrift für Operations Research. Serie A. Serie B*, 24(5):177–190, 1980.

[Irl02] Albrecht Irle. *Finanzmathematik.* Teubner, Wiesbaden, 2002.

[Irl05] Albrecht Irle. *Wahrscheinlichkeitstheorie und Statistik: Grundlagen - Resultate - Anwendungen.* Teubner Verlag, 2005.

[Irl06] Albrecht Irle. A forward algorithm for solving optimal stopping problems. *Journal of Applied Probability*, 43(1):102–113, 2006.

[Irl09] Albrecht Irle. On forward improvement iteration for stopping problems. In *Proceeding of the International Workshop of Sequential Methodologies*, Troyes, 2009.

[KS06] Anastasia Kolodko and John Schoenmakers. Iterative construction of the optimal Bermudan stopping time. *Finance and Stochastics*, 10(1):27–49, 2006.

[Nev75] Jacques Neveu. *Discrete-parameter martingales.* Elsevier, 1975.

[Pre10] E. L. Presman. A new approach to the solution of optimal stopping problem. In *Conference Proceedings*, Turku, 2010.

[PS73]   E. L. Presman and I. M. Sonin. The best choice problem for a random number of objects. *Theory of Probability and its Applications*, 17:657–668, 9 1973.

[Put94] Martin L. Puterman. *Markov Decision Processes: Discrete Stochastic Dynamic Programming*. Wiley-Interscience, 1994.

[Sne52] James Laurie Snell. Applications of martingale system theorems. *Transactions of the American Mathematical Society*, 73:293–312, 1952.

Die VDM Verlagsservicegesellschaft sucht für wissenschaftliche Verlage abgeschlossene und herausragende

Dissertationen, Habilitationen, Diplomarbeiten, Master Theses, Magisterarbeiten usw.

für die kostenlose Publikation als Fachbuch.

Sie verfügen über eine Arbeit, die hohen inhaltlichen und formalen Ansprüchen genügt, und haben Interesse an einer honorarvergüteten Publikation?

Dann senden Sie bitte erste Informationen über sich und Ihre Arbeit per Email an *info@vdm-vsg.de*.

Sie erhalten kurzfristig unser Feedback!

VDM Verlagsservicegesellschaft mbH
Dudweiler Landstr. 99
D - 66123 Saarbrücken

Telefon +49 681 3720 174
Fax   +49 681 3720 1749

www.vdm-vsg.de

Die VDM Verlagsservicegesellschaft mbH vertritt

Printed by Books on Demand GmbH, Norderstedt / Germany